Down to the Wire

Other Books by David W. Orr

Ecological Literacy (1992)

Earth in Mind (1994/2004)

The Nature of Design (2002)

The Last Refuge (2004)

Design on the Edge (2006)

The Global Predicament,
coedited with Marvin S. Soroos (1979)

The Campus and Environmental Responsibility,
coedited with David G. Eagan (1992)

The Sage Handbook of Environment and Society,
coedited with Jules Pretty et al. (2008)

Hope is an Imperative: The Essential David Orr (2010)

DAVID W. ORR

Down to the Wire

Wire

 *Confronting Climate Collapse*

OXFORD
UNIVERSITY PRESS

OXFORD
UNIVERSITY PRESS

Oxford University Press, Inc., publishes works that further
Oxford University's objective of excellence
in research, scholarship, and education.

Oxford   New York
Auckland   Cape Town   Dar es Salaam   Hong Kong   Karachi
Kuala Lumpur   Madrid   Melbourne   Mexico City   Nairobi
New Delhi   Shanghai   Taipei   Toronto

With offices in
Argentina   Austria   Brazil   Chile   Czech Republic   France   Greece
Guatemala   Hungary   Italy   Japan   Poland   Portugal   Singapore
South Korea   Switzerland   Thailand   Turkey   Ukraine   Vietnam

Copyright © 2009 by Oxford University Press, Inc.
First issued as an Oxford University Press paperback, 2012

Published by Oxford University Press, Inc.
198 Madison Avenue, New York, NY 10016

www.oup.com

Oxford is a registered trademark of Oxford University Press

Library of Congress Cataloging-in-Publication Data
Orr, David W.
Down to the wire : confronting climate collapse / David W. Orr.
    p. cm.
Includes bibliographical references and index.
ISBN 978-0-19-539353-8; 978-0-19-982936-1 (PBK.)
   1. Climatic changes. 2. Climatic changes—Environmental aspects.
I. Title.
QC981.8.C5O77   2009
363.34—dc22   2009005584

For my grandchildren Lewis, Molly, Ruby Kate, and Maggie.

Contents

Preface to the Paperback Edition

*"We seem oblivious to the danger—unaware how close we may
be to a situation in which a catastrophic slip becomes practically
unavoidable, a slip where we suddenly lose all control and are pulled
into a torrential stream that hurls us over a precipice to our demise."*
—James Hansen, 2009

I WROTE *DOWN TO THE WIRE* BETWEEN 2007 AND 2008,
when many still believed that the United States was capable of
making an effective national response to global climate destabilization. At the time I was involved with several dozen others in
drafting "The President's Climate Action Plan" (PCAP), a document that aimed to define the actions that the next U.S. President
would have to take immediately in order to avoid the worst of
what lies ahead. The numbers are stark. We have raised the temperature of the Earth by 0.8°C and are committed to another 0.5
to 1°C that will bring us close to the 2°C warming that many scientists have cited as a threshold we should not transgress.[1] Time, in

---

[1] The Royal Society (2010) report prior to the Cancun Conference of the
Parties includes a plausible estimate of a 4°C rise by 2061.

other words, is not on our side. Climate destabilization, we argued, was not just another issue on a long list of vexing problems, but the linchpin issue the solution of which would lessen many other problems including national security, balance of payments, economic recovery, and public health. As Bill Becker, the Executive Director of PCAP put it, "climate change is not about Right or Left; it's about backward or forward." Our efforts built on those of many others dating back to the early 1970s who pointed out that for many good reasons energy efficiency and deployment of renewable energy were the smartest, cheapest, and best policy options we have. The difference between the first warning about climate change, given to President Johnson in 1965, and November, 2008, when the PCAP report was delivered to John Podesta, Director of President Obama's transition team, was the accumulation of overwhelming scientific evidence that climate destabilization would most certainly threaten our security and well-being as a nation, and would, in time, erode the foundations of civilization itself. The scientific evidence is clear that the magnitude of the changes ahead are greater, the speed much faster, and duration of climatic destabilization longer-lasting than once thought.

The economic collapse of 2008 diverted attention from climate policy to economic recovery, yet they were, and remain, related problems. A robust, well-thought-out climate and energy policy would have amplified the effects of stimulus funding, which, by reliable accounts, was about half of what was actually needed. On the other hand, the economic crisis was "solved," but in ways that compounded the long-term climate problem. Indeed, a good fraction of stimulus funding went to projects that will both directly and indirectly increase U.S. carbon emissions.

This is not to say that the administration did nothing; President Obama launched the largest effort in U.S. history to deploy solar and wind power and raise national standards for energy efficiency. Behind the scenes, the administration has taken significant steps, much as recommended in the PCAP report, to redirect

Federal policies for purchasing, building standards, and research in advanced energy technologies. What the President did not do, however, was to use the power of the Office as a "bully pulpit" to lead public opinion when he still had the opportunity to do so. Nor did he make climate and energy policy the first priority of his administration. Instead, he chose to pursue a "bi-partisan" approach with an intransigent Republican minority that would have none of it, placed a politically divisive healthcare reform battle before the health of the planet, and let Congress patch together whatever legislation it could to deal with climate issues. Perhaps the President and his advisers reasoned that climate destabilization would not be all that bad despite the scientific evidence. Perhaps they thought that the issue could be deferred to a more convenient time. Maybe they believed that it would be politically impossible to do much anyway. In fact, by their failure to lead, they helped make it impossible—at least for a while. In short, whatever his many other successes, President Obama has so far failed the test of transformative leadership on what is likely to be the most important issue humankind will ever face.

In the absence of firm and decisive Presidential leadership on climate destabilization, among other factors, the two years since have not been our finest. The Waxman-Markey Bill (American Clean Energy and Security Act) with all of its complexities and compromises, passed the House of Representatives in 2009, but the U.S. Senate could muster no filibuster-proof majority on climate legislation, and had it done so, the results would have been woefully inadequate at best. We needed then, and need now more than ever, national climate and energy legislation that eliminates upwards of ninety percent of our greenhouse gas emissions and that is also fair, transparent, and straightforward. This needs to be executed with wartime urgency. Indeed, we are at war, but with ourselves and the devils of our lesser nature. At the international level, lacking vigorous U.S. leadership, climate conferences at Copenhagen (2009) and Cancun (2010) produced little progress relative to the magnitude of the looming crisis. In the meantime other countries are surging ahead to seize the lead and reap the economic advantages of a solar and wind powered world.

In the hardcover edition I reflected on the failures of U.S. politics; I have little to add, except to say it's even worse now than it was then. In Bill Becker's words: "climate change remains an issue whose name cannot be spoken, except in denial. Contrived scandals, deliberate misinformation, and scare tactics about higher energy prices and lost jobs have encouraged many politicians to treat climate change as a third-rail issue" (PCAP, 2011, 84). After the 5-4 Supreme Court ruling in *Citizens United* (2009), the power of secret and unaccountable money is more pervasive than ever before. The "Tea Party" movement, funded by those with much to gain from public befuddlement, misdirected anger, and more deregulation, changed the political landscape—at least for the time being—and effectively closed any window of opportunity for Federal climate legislation that existed in early 2009. Too much money in politics, the power of an unaccountable extreme right-wing media, failed leadership (including first and foremost that of the Republican party), and too much power in the hands of Senators representing more acres and cows than people add up to what Eric Alterman (2011) has called "Kubuki Democracy"— a system that is rigged to prevent solutions to public problems and seemingly incapable of repairing itself. And there is, as Peter Burnell put it, "no certainty that democracies will do all that is required and in a timely fashion" anyway (2009, p. 36). But beneath the surface of U.S. politics in particular are deeply rooted beliefs that individual liberty trumps the public good and that government is most always wrong except when it endeavors to wither away—a belief with strange echoes in Marxism.

The capacity and apparent willingness of humankind to destabilize the climate conditions that made civilization possible is *the* issue of our time; all others pale by comparison. Beyond some unknown threshold of irreversible and irrevocable changes driven by carbon cycle feedbacks, climate destabilization will lead to a war of all against all, a brutal scramble for food, water, dry land, and safety (Dyer, 2010). Sheer survival will outweigh every other consideration of decency, order, and mutual sympathy. Climate destabilization will amplify other problems caused by population growth, global poverty, the spread of weapons of mass destruction, and the potential impact of high consequence events that have long-term global consequences—what risk analyst Nassim Taleb calls "black swans."

From this, two large questions arise. First, how do we think about the issue of climate destabilization? Second, what do we do about it? In polite circles, the first question is seldom asked because it runs counter to the faith that climate destabilization is merely another problem to be solved by better technology and proper market signals. I hasten to say that I am all for better technology and markets that accurately price things of true value—things like climate stability and the well-being of our descendents. But climate destabilization is more than a technical or economic issue. We ought to ask why we are coming so close to the brink of global disaster so casually and carelessly. We ought to ask why the market—skewed to the advantage of corporations and the super wealthy—is allowed to trump the rights of our descendents to "life, liberty, and property" which presupposes climate stability. We ought to ask why the ideology of markets has so thoroughly saturated our language, thought, morality, and behavior so that we can hardly imagine any other standard for human conduct and national policy beyond short-term pecuniary advantage. We ought to ask why our leaders remain so deferential to predatory bankers, financiers, and corporate tycoons who nearly collapsed the global economy in 2008. We ought to ask why human nature, or just the nature of the American sub-species, is so vulnerable to the "merchants of denial" (Naomi Oreskes and Conway, 2010; Weart, 2011). Answers to such questions would get us close to the roots of the problem, which is found in the unholy alliance between corporations, the communications industry, government, the military, and security organizations much as Dwight Eisenhower warned in his 1961 farewell address.

How we think about climate destabilization has a great deal to do with how we talk about it. For example, we do not face a mere "warming" of the Earth, but rather a worsening destabilization of almost everything. We are rapidly making a different and less hospitable planet, one that writer and activist Bill McKibben calls "Eaarth" (2010). But as ethicist Clive Hamilton wrote, "The language now used—'risk management', 'adapting', 'building resilience', 'no-regrets', 'win-win'—reflects the belief that to accommodate a warmer world we need only tinker at the edges of the system." (Hamilton, 2010, p. 206). Moreover, among deniers, the issue is often cast as a matter of belief. But climate science is not a matter of faith; its physical laws are subject to rigorous and replicable proof. By comparison, few people say things like "I don't believe in the law of gravity", for which there are simple and fairly definitive tests, all of which end abruptly. Climate deniers have a double standard. Confronting a life-threatening illness, they will presumably seek medical help from physicians scientifically trained to understand the body, and thereby improve their chances of survival. But warned of irrevocable planetary disaster ahead by real scientists who must live by the rigors of peer review, fact, data, and logic deniers retreat down the rabbit hole into a science-free zone where things are so because they are said to be so.

Because the issue is unlike any we have ever faced before, it would be difficult enough to handle without deliberate distortion and outright lies. The consequences are global and, beyond some threshold, they will be irreversible and catastrophic. The science is complex, as are the myriad of related social and ethical aspects. Climate and energy issues divide the wealthy against the poor, nation against nation, and the present generation against its own posterity. Furthermore, the worst effects are still hidden from view—the possibility of rapid climate destabilization invites procrastination and denial. Anything like a proper response at this late hour will require unprecedented wisdom and a manner of comprehensive thinking and acting at a scale and immediacy for which we have no good examples. It is indeed the "perfect problem," or what in policy circles are known as "wicked problems" (Balint, et.al., 2011), yet we continue to talk about climate destabilization as if it were an ordinary issue requiring no great vision, no unshakable resolve, no fear of the abyss.

Instead, many continue to believe that our failure to respond adequately is the result of our failure to present a positive image. We have, they assert, marinated too long in "doom and gloom." Their advice, instead, is to be cheery, upbeat, and talk of happy things like green jobs and more economic growth, but whisper not a word about the prospects ahead or the suffering and death already happening. Perhaps that is a good strategy and there is room for honest disagreement. But "happy talk" (Gus Speth's phrase) was not the approach taken by Lincoln confronting slavery, or by Franklin Roosevelt facing the grim realities after Pearl Harbor. Nor was it Winston Churchill's message to the British people at the height of the London blitz. Instead, in these and similar cases, transformative leaders told the truth honestly, with conviction and eloquence.

I believe that the same standard should apply to us. We must have the courage to speak the truth, and the vision and fortitude to chart a plausible way forward. The truth of the matter is that even in the best scenarios imaginable, we would still have a long and difficult road ahead before climate stabilizes again, hopefully within a range still hospitable to us. It is also true that we have the capability to make the transition to economies powered by sunlight and efficiency. The point is not to be gloomy or cheery, but to be truthful and get to work.

With no prospect for Federal climate legislation anytime soon, however, what's to be done? The short answer is that, whatever the prospects, we must keep pushing on every front to change Federal and state policies, transform the economy, improve public understanding of science, engage churches and civic organizations, reform private institutions, educate for ecological literacy, and change our own behavior. Even without a coordinated, systematic national response, maybe enough small victories in time will suffice.

I ended *Down to the Wire* on a personal note by describing the Oberlin Project, a joint effort by Oberlin College and the City of Oberlin. Our goals include rebuilding a 13- acre block in the downtown area to the U.S. Green Building Platinum Standard as a driver for economic revitalization, making the transition to carbon neutrality, developing a 20,000 acre greenbelt, and completing all of the above as a part of an educational venture that joins the public schools, the college, a community college, and a vocational educational school. We aim to equip young people for decent and creative lives in a post-cheap fossil fuel economy. The Project, now two years in the making, is organized around ten community teams working on strategic issues such as energy, public policy, finance, community, economic development, and education. We aim to develop what Patrick Doherty of the New America Foundation has called "full-spectrum sustainability," in which the parts reinforce the resilience and prosperity of the entire community. In plain words, that means lots of meetings between different teams and stakeholders, because applied sustainability crosses virtually all of the fences and walls by which we organized the industrial world.

Like light refracted through a crystal, the Project appears different from different vantage points. For Oberlin students it means a cool 24/7 downtown. For college faculty it means better facilities. To the local merchants it means more business and higher profits. To public officials it is a model of climate neutral economic revitalization in a region devastated by de-industrialization. To architects and urban planners, it is a model of ecological design at the scale of a small city. To educators it is a model of applied pedagogy and hands-on learning. Over the past year I've come to believe that it is also a model of planning and development that will be necessary everywhere for reasons of security broadly defined (Matthews, 1989; Orr, 2011).

Local security—by which I mean safety *and* access to food, water, energy, shelter, health, and livelihood—was once assumed to be synonymous with our national capacity to project military power beyond our shores and borders. As such, security had little or nothing to do with how we designed, managed, and maintained the environmental and energy infrastructure of the country, or with the protection of its air, waters, soils, landscapes, biological diversity, or public health. But security in the twenty-first century will be a far more complicated and difficult challenge. In addition to traditional security challenges, we must now reckon with:

- terrorist threats to critical infrastructure—notably the electric grid (2007 Defense Science Board);
- food shortages, water scarcity, and expensive oil by 2030 or sooner—a problem described by the chief science advisor to the British government, John Beddington, (2009) as a "perfect global storm";
- the implications of rapid climate destabilization, including mass migrations estimated to reach 250 million by midcentury. The massive heat wave of 2010 in central Russia, record heat in Asia, unprecedented flooding in Pakistan, and the largest cyclone in Australian history (February, 2011) are consistent with projections by most climate scientists, and likely portents of what's ahead for all of us; and
- increasingly frequent low probability/high global consequence events such as the financial crisis of 2008, or technological accidents, infrastructure failures, and acts of God (Taleb, 2010).

The upshot, in Joshua Cooper Ramo's words, is that "we must squarely face the awful fact that our security will become ever more perilous." (2009, p. 97)

We must also face the fact that no government on its own can protect its people in a black swan world or from the growing impacts of climate destabilization and the turmoil likely to accompany the transition to a post-fossil fuel world. Citizens, neighborhoods, communities, towns, cities, regions, and corporations will have to do far more to ensure reliable access to food, energy, clean water, shelter, and economic development in the decades ahead. We are reaching the political, organizational, and ecological limits of large-scale enterprises, whether governmental or corporate, to be the sole-source providers for a largely passive and dependent public. This is not to argue against Federal policy changes to promote sustainable development, reform the tax system, deploy clean energy, and improve public transportation—things that can best be done or only done by the Federal government. But the reality is that communities will have to carry much more of the burden than heretofore. National security and local security, in other words, are now joined as a part of a larger narrative in which considerations of security, climate policy, and sustainable development must be joined.

Sustainability, in short, must be the domestic and strategic imperative for the twenty-first century. Its chief characteristic is resilience—a concept long familiar to engineers, mathematicians, ecologists, designers, and military planners—which means the capacity of the system to "absorb disturbance; to undergo change and still retain essentially the same function, structure, and feedbacks" (Walker and Salt, 2006, p. 32; Lovins and Lovins, 1983, chapter 13; Lovins, 2002). Resilient systems are characterized by redundancy so that failure of any one component does not cause the entire system to crash. They consist of diverse components that are easily repairable, widely distributed, cheap, locally supplied, durable, and loosely coupled. In Joshua Ramo's words: "studies of food webs or trade networks, electrical systems and stock markets, find that as they become more densely linked they also become *less* resilient; networks, after all, propagate and even amplify disturbances" (Ramo, 198). In practical terms, resilience is a design strategy that aims to reduce vulnerabilities, often by shortening supply lines, improving redundancy in critical areas, bolstering local capacity, and solving for a deeper pattern of dependence and disability. The less resilient the country, the more military power is needed to protect its far-flung interests and client states, hence the greater the likelihood of wars fought for oil, water, food, and materials. But resilient societies need not send their young to fight and die in far-away battlefields, nor do they need to heat themselves into oblivion.

While we have become more vulnerable to a wide range of threats, a revolution in the design of resilient systems has been quietly building momentum for nearly half a century. It includes dramatic changes in:

- architecture, e.g. buildings and communities powered entirely by efficiency and renewable energy;
- waste management in which all wastes are purified by natural processes;
- agriculture that mimics natural systems;
- renewable energy technologies;
- advances in energy efficiency;
- cradle to cradle and biomimetic production systems that create no waste;
- urban planning and smart growth strategies that build ecologically coherent cities; and
- tools for systems analysis that improve foresight, organizational learning, and policy integration.

These and other advances in science, technology, and policy are the tools and technologies for a society that is more secure by design, and hence more resilient in the face of disruptions, whether by malice, climate change, accidents, human error, or acts of God. They are the necessary foundation for policies that are less provocative to other nations and less likely to engender global conflicts while reducing balance of payments deficits for imported oil and thereby:

- Eliminating our dependence on politically unstable regions;
- Cutting military costs associated with oil dependence;
- Eliminating our carbon emissions;
- Equipping the next generation for lives and livelihoods in economies and societies calibrated to work with natural systems;
- Increasing our prosperity by creating employment and business in sustainable enterprises; and
- Improving the capacity of communities and regions to withstand effects of climate destabilization and new threats to critical infrastructure as well as global economic turmoil.

Said differently, security is too important to be left solely to the generals, defense contractors, and politicians in Washington. It will be necessary for neighborhoods, communities, towns, cities, and regions to improve their resilience and security by their own initiative, ingenuity, and foresight. The Oberlin Project is one example, but there are many others at different scales and in different regions. We have begun to join many of these in a "National Sustainable Communities Coalition security network of sustainability sites, cities, and projects" which aims to improve local and regional resilience to plausible future circumstances. Remove the word "national" and imagine a global network of transition towns (Hopkins, 2007), cities, regions, and organizations—a solar powered renaissance of local capability, independence, culture, and security in the full sense. Imagine a world, someday, where no child need fear violence, hunger, thirst, poverty, ignorance, homelessness, or heat and storms beyond imagining.

Preface

The ongoing disruption of the Earth's climate by man-made green-house gases is already well beyond dangerous and is careening toward completely unmanageable.

—John Holdren

ALL OF US BREATHE FROM THE SAME ATMOSPHERE, DRINK the same waters, and are fed from the land. All of us depend, more than we can know, on the stability of the same biogeochemical cycles, including the movement of carbon from plants to the atmosphere, oceans, soils, and living creatures. All of us are vulnerable to the remorseless workings of the large numbers that govern Earth systems. All of us are stitched to a common fabric of life, kin to all other life forms. All of us are products of the same evolutionary forces and carry the marks of our long journey in time. Each of us is a small part of a common story that began three billion years ago. We are all made of stuff that was once part of stars, and we will all become dust to be remade someday into other life forms. As persons, we are visitors on the Earth for only a brief moment. As a species, however, we are in our adolescence, and as is common at that stage of life we live dangerously. Specifically, we have created three ways to commit suicide: by nuclear annihilation, by ecological degradation, and, as computer scientist Bill Joy notes, by the consequences of our own cleverness—eviction by technologies that can self-replicate and might one day find *Homo sapiens* useless and inconvenient.[1] We have entered an era that Harvard biologist Edward O. Wilson calls "the bottleneck" (Wilson, 2002, pp. 22–41).

This book is written in the belief that we will come through that gauntlet chastened but improved. But it will be trial by fire, hopefully, a tempering process in which we will shed our illusions of being separate from nature and our pretense that we can master nature or each other through violence. On the other side of the bottleneck, maybe we will have gained a clearer vision of the value of life and a deeper understanding of what it means to be stewards and trustees for all life to come. But this is certainly not the only scenario one might imagine—perhaps, it is not even very likely. There are darker possibilities with which we must contend and which we must have the foresight to anticipate and the wisdom to avoid.

In the fossil fuel age we lived in the unspoken faith that there are no "booby traps for unwary species," as biologist Robert Sinsheimer once put it. Unwittingly we set our own, and now the carbon trap is nearly sprung. Even before the coal and oil age we exploited carbon-rich soils and forests, and that is the history of rising and falling empires and the uneven march of civilization. The trap was founded on ignorance of our impact on the biogeochemical cycles of Earth, which posed no serious problems when we were fewer and depended on sunlight and wind for our energy. But now the six and a half billion of us, soon perhaps to be eight or nine billion, are living carbon-intensive lives. We set the trap and it will now take our most creative and sustained efforts to avert catastrophe, and that will require reducing our carbon footprint from 22 tons per person per year to 1–2 tons or even less. But even then, "when this centuries-long climate storm subsides, it will leave behind a new, warmer climate state that will persist for thousands of years. That's the basic outlook" (Archer, 2009, p. 45).

Even in the near term it is already too late, however, to avert significant disasters, and that is a difficult message to convey without inducing paralysis or denial even among those willing to listen. It is a great deal easier for all of us to hit the snooze button on the alarm clock, go back to sleep, and hope that it all goes away, or to pretend that dire circumstances present only opportunities. Climate change presents opportunities for some, certainly, but for the Tuvalu islanders, the victims of floods and droughts and of larger hurricanes and typhoons, those living in low-lying areas like Bangladesh, and the 150,000 who die each year in climate change–driven weather events, the word "opportunity" has a peculiarly hollow sound. It will as well for the 250,000,000 or more climate refugees that the United Nations estimates will be homeless by midcentury.

Through the coming decades and centuries of the bottleneck, great leadership at all levels will be essential. We will need leaders first, with the courage to help the public understand and face what will be increasingly difficult circumstances. The primary cause is climate destabilization, described in four consensus reports by the Intergovernmental Panel on Climate Change over 20 years and hundreds of other scientific reports. Often, however, we dismiss bearers of bad news or inconvenient truths until the point of crisis, when reality can no longer be evaded. The mythical figure of Cassandra and the Old Testament prophet Jeremiah were fated to be ignored until it was too late to avoid the dire things they foretold. The same disbelief has greeted the increasingly frequent and rigorous warnings in our time. One of the earliest, for example, was issued by the Council on Environmental Quality in the Carter administration and published in 1980 as the

Global 2000 Report (Barney, 1980). The authors catalogued in great detail the scientific evidence about declining ecosystems, climate change, and species loss, along with measures necessary to move the country toward sustainability. But we chose to evade reality and sought refuge in the slogan that it was "morning in America again." Three decades later it is twilight, and we live with the ecological, economic, political, and social consequences of our own making.

Second, in the "long emergency"[2] ahead leaders will need an uncommon clarity about our best economic and energy options. Some choices being proposed by well-funded and highly organized lobbies would commit the nation and the world to courses of action that will lead to unfortunate and irreversible consequences. They will need to understand their relative costs, risks, and benefits, including those over the long term, to avoid making decisions that lock us in to policies that we—or our children—will someday sorely regret. There are better possibilities that would go a long way toward solving the underlying causes of our problems. But knowing which is which requires that they recognize the difference between the structure of problems and their coefficients—the rate at which they get worse. In other words, they need to understand the difference between Band-Aids and authentic cures, and that requires that we better understand otherwise obscure concepts like feedback loops, leads, and lags, which is to say how the world works as a unified system (Meadows, 2008). They must see, in other words, the many connections between climate, environment, prosperity, security, and fairness. In this perspective, climate destabilization is not an aberration but a predictable outcome of a system haphazardly created in the dim light of a dangerously incomplete image of reality.

The results are increasingly clear: even were we to stop emitting heat-trapping gases quickly, we will still experience centuries of bigger storms, larger and more frequent floods, massive heat waves, and prolonged drought, along with rising sea levels, disappearing species, changing diseases, decline of oceans, and radically altered ecosystems.³ In the long emergency ahead, people, communities, societies, institutions, organizations, and global society will be sorely stressed. The third quality of leadership in these circumstances is the capacity to foster a vision of a humane and decent future. Such a future will require a great deal of kindness for growing numbers of people who will need our help as friends, neighbors, community members, and fellow sojourners on this fragile craft that we call civilization. Eventually, we will need their help as well. No one will remain unaffected by climate destabilization and its many consequences that will spill across the boundaries of geography, circumstance, and time.

The news about climate, oceans, species, and all of the collateral human consequences will get a great deal worse for a long time before it gets better. The reasons for authentic hope are on a farther horizon, centuries ahead when we have managed to stabilize the carbon cycle and reduce carbon levels close to their preindustrial levels, stopped the hemorrhaging of life on Earth, restored the chemical balance of the oceans, and created governments and economies calibrated to the realities of the biosphere and to the diminished ecologies of the postcarbon world. The change in our perspective from the nearer to the longer term is, I think, the most difficult challenge we will face. We have become a culture predicated on fast results, quick payoffs, and instant gratification. But now we will have to summon the fortitude necessary to undertake a longer and more arduous journey. Rather like the builders of the great cathedrals of Europe, we will need stamina and faith to work knowing that we will not live to see the results.

I begin by assuming the most optimistic outcome possible—that, by a combination of advanced technology and wise policy choices, the world will quickly act to stabilize concentrations of greenhouse gases and reduce emissions to a level below that which would lead to runaway climate change. Nonetheless, barring some quite unexpected technological breakthrough, the consequences of what we have already "bought" will still cause great hardship everywhere. Glib talk about "climate solutions" misleads by conveying the impression that climate is merely a problem that can be quickly solved by technological fixes without addressing the larger structure of ideas, philosophies, assumptions, and paradigms that have brought us to the brink of irreversible disaster. The point is the same as one that has been attributed to Einstein: "significant problems we face cannot be solved at the same level of thinking we were at when we created them" (Calaprice, 2005, p. 292). There are certainly better technologies to be deployed, and far better ones soon to come. But the climate is not likely to be restabilized by any known technical fix quickly, easily, or painlessly. Rather, as geophysicist David Archer puts it:

> The climatic impacts of releasing fossil fuel CO_2 to the atmosphere will last longer than Stonehenge. Longer than time capsules, longer than nuclear waste, far longer than the age of civilization so far... [it] will persist for hundreds of thousands of years into the future. (Archer, 2009, pp. 1, 90)

Climate change, in other words, is not so much a problem to be fixed but rather a steadily worsening condition with which we must contend for a long time to come. Improved technology, at best, will only reduce the scale of the problem and buy us time to build the foundations for a more durable and decent civilization. In the words of biologist Anthony Barnoski, "stabilizing [climate] in this sense means global temperature staying more or less constant for at least hundreds, probably thousands of years. In short, as far as generations of humans are concerned, we probably never will revert back to the 'old' climate" (2009, p. 29).

The few remaining climate skeptics aside, there are two general positions that bear on my own views. The first is the belief that there is a rising tide of groups, associations, and nongovernmental organizations forming around the world as a kind of planetary immune system that will transform our politics, heal the widening breach between humankind and the rest of nature, and lead on to sunnier uplands. There is considerable evidence for what Paul Hawken calls "blessed unrest." Clearly something is astir in the world, and perhaps it will eventually transform our manner of living and relating to the world and to each other. But it has not done so yet. In the meantime, carbon is accumulating in the atmosphere faster than ever before while inequality, violence, economic stress, and ecological degradation grow. How blessed unrest amplified by the Internet will fare in an increasingly destabilized world is anyone's guess, but to get through the bottleneck more or less intact we will need lots more of it, well organized, creatively applied, and allied with leadership in all sectors of society. But there is no adequate substitute for better leadership at all levels, including those who are engaged in the conduct of the public business, which is to say politics.

A second view holds that we ought to focus only on solutions, not problems and dilemmas. But the solutions most talked about are technological and so neither require nor result in any particular improvement in our behavior, politics, or economics that brought us to our present situation in the first place. And neither do they call us to rethink the rationality of our underlying motives and objectives or become aware of the political and social choices hidden in our technologies (Winner, 1986, pp. 19–39). The aim, merely, is to do what we are already doing more efficiently and effectively without asking whether it is worth doing at all. We ought, it is said, to make hope possible, not despair plausible. I believe that to be a good rule until wishful thinking masquerades as hope and avoidance of despair becomes evasion of reality. Those who focus exclusively on solutions are rather like doctors who only prescribe and never diagnose. In the real world an effective prescription depends a great deal on an accurate diagnosis of the nature and source of the problem. After decades of hyperconsumerism and worship of commerce, a dose of reality, with or without despair, would lay the foundation for a more grounded, sober, and authentic hope. Our best chance of surviving through the long emergency ahead lies in our capacity to face difficult facts squarely, think clearly about our possibilities, and get down to work.

The faith placed in better technology is tied to the faith in unfettered markets and commerce, the reputation of which had been much improved due to the efforts of Milton Friedman and his free-market disciples until the economic collapse of 2008. The appeal to economic self-interest as the engine of human progress has its origins in the writings of Adam Smith, and there is much to be said on its behalf. Forgotten in the euphoria, however, are Smith's own misgivings about the results of unalloyed self-interest, evident in both *The Wealth of Nations* and *The Theory of Moral Sentiments*. Until the great financial implosion of 2008, amnesia also veiled the spotty and often shabby record of corporations and financial institutions operating without the countervailing power of alert governments and an engaged and sometimes enraged citizenry. Economists, nonetheless, are inclined to attribute all societal shortcomings to a failure of markets, and sometimes, in some ways, they are. But the belief that climate destabilization represents "the largest market failure in history" is misleading because it overlooks a prior and larger failure of political leaders to acknowledge the problem before it grew into a crisis. Even with ample and increasingly urgent warnings, they failed to restructure the rules and regulations that govern the use of fossil fuels when it would have been relatively easy and cheap to minimize or avoid much of the crisis altogether.

I write, accordingly, as an advocate for better leadership, an improved democracy in the United States, and more creative and competent management of the public business. Climate destabilization is obviously a global crisis, but I've chosen to narrow my focus to the United States because we are the largest economy on Earth and the largest source of heat-trapping gases in the industrial era, and we have greater leverage on the issue than any other country. And for no good reason we were absent without leave until very recently on the largest issue ever on the human agenda. The United States, in other words, is not just another country; it is, rather, the linchpin in the effort to avoid catastrophic global destabilization.

Finally, this book is a companion of sorts to a project launched in June of 2006 at a Wingspread conference convened by Ray Anderson, Bill Becker, and Jonathan Lash, members of President Clinton's Council on Sustainable Development, which had gone dormant in the years of George W. Bush. Among the recommendations from that conference was one I made to create a climate action plan for the first hundred days of the next U.S. president.[4] The idea was accepted and funded by Adam Joseph Lewis, the Rockefeller Brothers Fund, and others. The project was cochaired by Ray Anderson and Gary Hart and ably directed by Bill Becker. The final report, presented to John Podesta, director of the Obama transition team, included some three hundred proposals across a dozen categories ranging from transportation to land use. That document was aimed at near-term specific policy changes—the things the next U.S. president and the government would have to do quickly to respond to the challenge of climate destabilization. This book, by contrast, addresses the larger issues behind the immediate policy choices and headlines. It is a meditation on the leadership we will need to eventually surmount the largest challenges we've ever experienced. My focus is what historian James MacGregor Burns describes as transformational leadership that recognizes "real need, the uncovering and exploiting of contradictions among values and between values and practice, the realigning of values, reorganization of institutions where necessary, and the governance of change. Essentially the leader's task is consciousness-raising on a wide plane" (Burns, 1978, p. 43). And we will need a great deal of consciousness-raising in the years ahead.

Acknowledgments

THIS BOOK OWES MUCH TO THE FRIENDSHIP, COLLABORA-
TION, and counsel of many over many years and to others I have
not met but who have instructed through their example, writ-
ings, and leadership. My thanks to John Adams, Paul Alsenus, Ray
Anderson, Kenny Ausubel, Zenobia Barlow, Taylor Barnhill, Andy
Barnett, Seaton Baxter, David Beach, Bill Becker, Frances Bei-
necke, Janine Benyus, Bob Berkebile, Scott Bernstein, Thomas
Berry, Wendell Berry, Rosina Bierbaum, Jessica Boehland, Nina
Leopold Bradley, Lester Brown, Peter Brown, Bill Browning, Peter
Buckley, Fritjof Capra, Majora Carter, Rick Clugston, Leila Con-
nors, Peter Corcoran, Tony Cortese, Bob Costanza, David Crock-
ett, Michael Crow, John Curry, Herman Daly, Leo DiCaprio,
Marcellino Echeverria, David Ehrenfeld, Jim Elder, John Elder,
Richard Falk, Chris Flavin, Karen Florini, Peter Forbes, Eric Frey-
fogle, Howard Frumkin, Ross Gelbspan, Larry Gibson, Marion
Gilliam, Teddy Goldsmith, Zac Goldsmith, Eban Goodstein, Al
Gore, John Grim, Maria Gunnoe, Bruce Hannon, Jim Hansen,
Gary Hart, Nick Hart-Williams, Paul Hawken, Denis Hays, Teresa
Heinz, Mary Anne Hitt, John Huey, Buddy Huffaker, Wes Jack-
son, Sadhu Johnston, Van Jones, Greg Kats, Steve Kellert, Julian

Keniry, Robert Kennedy, Bob Kerr, David Kline, Bob Koester, Fred Krupp, Satish Kumar, Jeremy Leggett, Carl Leopold, Estella Leopold, Adam Lewis, Gene Logsdon, Rich Louv, Tom Lovejoy, Amory Lovins, Hunter Lovins, Wangari Maathai, Arjun Makhijani, Ed Mazria, Carl McDaniel, Jay McDaniel, Bill McDonough, Bill McKibben, Gary Meffe, George Monbiot, Bill Moomaw, Kathleen Dean Moore, Bill Moyers, Wil Orr, Jon Patz, Matt Peterson, Michael Pollan, Carl Pope, Jonathan Porritt, John Powers, Jules Pretty, Steven Rockefeller, Kirk Sale, Chuck Savitt, Jonathan Schell, Stephen Schneider, Larry Schweiger, Peter Senge, Nina Simon, Robert Socolow, David Shi, Gus Speth, Paul Stamets, Frederick Steiner, Steven Strong, Bill Sullivan, Woody Tasch, Bill Thompson, John and Nancy Todd, Mitch Tomashow, Mary Evelyn Tucker, Sim Van der Ryn, Steve Viederman, Bill Vitek, Mathis Wackernagel, Greg Watson, Burns Weston, Bob Wilkinson, Alex Wilson, Edward O. Wilson, George Woodwell, and many others. Directly or indirectly, each has influenced my thinking about climate change and the proper role of humankind in the community of life. And to each of you, for your example, work, insight, and heroism, I am grateful.

My friend and colleague Steve Mayer was a patient, thoughtful, and perceptive sounding board for many of the ideas in the book. I am grateful as well to other colleagues at Oberlin College, including David Benzing, Bev Burgess, Norman Craig, President Marvin Krislov, Roger Laushman, Bob Longsworth, Jane Mathison, Carl McDaniel, Tom Newlin, John Petersen, Richard Riley, Rumi Shamin, Harlan Wilson, and Cheryl Wolfe.

Thanks to Todd Baldwin, Stephen Dodson, David Ehrenfeld, Neva Goodwin, Tom Lovejoy, and Tisse Takagi for helpful comments on the manuscript. A special thanks to Peter Prescott for his encouragement, diligence, editorial skill, and friendship.

And to Elaine, Mike, and Dan for more than words can say.

Down to the Wire

Introduction

There were rumors of unfathomable things, and because we couldn't fathom them we failed to believe them, until we had no choice and it was too late.

Nicole Krauss, *The History of Love*

IN OUR FINAL HOUR (2003), CAMBRIDGE UNIVERSITY astronomer Martin Rees concluded that the odds of global civilization surviving to the year 2100 are no better than one in two.[1] His assessment of threats to humankind ranging from climate change to a collision of Earth with an asteroid received good reviews in the science press, but not a peep from any political leader and scant notice from the media. Compare that nonresponse to a hypothetical story reporting, say, that the president had had an affair. The blow-dried electronic pundits, along with politicians of all kinds, would have spared no effort to expose and analyze the situation down to parts per million. But Rees's was only one of many credible and well-documented warnings from scientists going back decades, including the Fourth Assessment Report from the Intergovernmental Panel on Climate Change (2007). All were greeted with varying levels of denial, indifference, and misinterpretation, or were simply ignored altogether. It is said to be a crime to cause panic in a crowded theater by yelling "fire" without cause, but is it less criminal not to warn people when the theater is indeed burning?

My starting point is the oddly tepid response by U.S. leaders at virtually all levels to global warming, more accurately described as "global destabilization." I will be as optimistic as a careful reading of the evidence permits and assume that leaders will rouse themselves to act in time to stabilize and then reduce concentrations of greenhouse gases below the level at which we lose control of the climate altogether by the effects of what scientists call "positive carbon cycle feedbacks."[2,3] Even so, with a warming approaching or above 2°C we will not escape severe social, economic, and political trauma. In an e-mail to the author on November 19, 2007, ecologist and founder of the Woods Hole Research Center George Woodwell puts it this way:

> There is an unfortunate fiction abroad that if we can hold the temperature rise to 2 or 3 degrees C we can accommodate the changes. The proposition is the worst of wishful thinking. At present temperatures, which would drift upward if the atmospheric burden were stabilized now, we are watching the melting of glaciers, frozen soil, and the accelerated decay of large organic stores of carbon in soils but especially in high latitude soils and tundra peat. A 2 degree average rise in global temperature will be 4–6 degrees or more in high latitudes, enough to trigger the release of potentially massive additional quantities of carbon dioxide and methane [that] would push the issue of control well beyond human reach.

John Podesta and Peter Ogden at the Center for American Progress concur, saying that even in the most optimistic scenario imaginable, "There is no foreseeable political or technological solution that will enable us to avert many of the climatic impacts projected" (Podesta and Ogden, 2008, p. 97).

The scientific evidence indicates that we have so far warmed the Earth by 0.8°C, and even if we were to suddenly stop emitting heat-trapping gases we would still be committed to another 0.5° to 1.0°C of warming, bringing us close to what many climate scientists regard as a dangerous threshold of 2°C above the preindustrial level. At some unknown level of human "forcing" of climate, however, further positive carbon cycle feedbacks will kick in and climate change will become a kind of runaway train.[4] A great deal depends on the sensitivity of the climate to human provocation, but no one can say for certain what margin of safety we have or whether we might have already transgressed that line.[5] What is known is that even without human forcing, "nonlinear, abrupt changes appear to be the norm, not the exception, in the functioning of the Earth system" (Steffen et al., "Abrupt Changes," 2004, p. 8).

Large and permanent risks to Earth notwithstanding, the use of fossil fuels continues to grow worldwide. The accumulation of carbon in the atmosphere is still accelerating, while some evidence suggests that sinks for carbon are decreasing. U.S. and Chinese emissions, in particular, continue to increase rapidly (Raupach et al., 2007). The roughly 30-year lag between the emission of CO_2 and its effects on climate means that the rapid melting of ice caps and glaciers, more severe droughts, heat waves, and storms visible today are the results of the fuels that we burned decades ago. In the meantime we have roughly doubled the flow of carbon into the atmosphere, and as a result are committed to a substantial further temperature increase. This is not just "global warming," however, but rather a progressive and accelerating destabilization of the entire planet. Some of the changes are predictable, but because of the complexity of the Earth and our ignorance of the full effects of various levels of forcing on the biosphere, others will come as nasty surprises. Changes are already apparent: spring comes earlier and winter arrives later, birds characteristic of southern regions are showing up in the north, storms and heat waves are more frequent and more severe. Around the globe new records for extreme weather are being set at a record-breaking pace. With another degree or so of warming, the changes will be unmistakable: traditional northern winters will be mostly a memory, food prices will rise sharply, forest fires will be more frequent, and many species will disappear. Maple syrup will no longer be made in Vermont. With still another degree, coastal cities like New Orleans, Miami, and Baltimore will eventually be flooded, the Everglades will disappear, Appalachian forests will be replaced by scrub trees and grasses, and a great human migration away from coasts and mid-continental regions will have begun[6] (Lynas, 2007). By then we will have created what climate scientist James Hansen describes as a "different planet," one we won't like. The upshot is that we now have every reason to believe, as scientist Wallace Broecker once put it, that the climate system is "an angry beast, and we are poking it with sticks" (Linden, 1997). We are now in a close race between our capacity to change at a global scale and the forces that we have unleashed.

Climate change, like the threat of nuclear annihilation, puts all that humanity has struggled to achieve—our cultures, art, music, literatures, cities, institutions, customs, religions, and histories, as well as our posterity—at risk. Unless we are led to act rapidly and wisely, we are on a course leading to an Earth of greatly reduced biological diversity populated by remnants and ruins. Had we acted sooner we would have had a far easier path and would have saved much more. But now problems are becoming a planetary crisis brought on by our own relentless growth, which affects the large numbers that govern the biosphere.

As the evidence mounted over the last three decades, the political response nonetheless was a combination of denial and delay. Confronted with evidence of the growing risks of planetary destabilization, many in positions of influence in government, media, business, the academy, and the far right of U.S. politics ignored and then later denied the facts. When the facts could no longer be denied, they quibbled about the details of the scientific evidence and the costs of action necessary to head off the worst possibilities. In the meantime, months, years, and decades slipped away. Some chose to dismiss the evidence in its entirety as "doom and gloom," but as individuals they lived by an entirely different calculus. They have household, auto, and health insurance for protection against vastly smaller risks at an infinitesimally smaller scale, and most did not dismiss health warnings from their doctors as a liberal plot. When it is merely the future of the Earth, however, they have been willing to risk irrevocable and irreversible changes.

On the positive side, polls indicate that public awareness about climate change is increasing rapidly. After years of inaction and denial, a new president of the United States supports serious action on climate change. Markets for carbon are coming into existence. Large amounts of capital are shifting toward low-carbon investments. Deployment of solar and wind energy is advancing rapidly worldwide. Billionaires like T. Boone Pickens are investing heavily in wind farms, not necessarily to save the Earth but to make money. Promising technologies are emerging from laboratories. And nongovernmental organizations, colleges and universities, and corporations are shifting priorities to accommodate and facilitate low- or zero-carbon futures. Led by California and Florida, states and regional coalitions are creating climate policy innovations. Hundreds of cities and local governments are developing policies to reduce carbon emissions and to adapt to changing climate. Hundreds of college and university presidents have committed to "climate neutrality." Polls show that the public is awakening and becoming increasingly supportive of action on climate change, energy efficiency, and solar power. A revolution has begun. There is a great temptation to stop here and accommodate the desire for happy news and the hope that we will not have to sacrifice economic growth, convenience, or comfort to avoid the worst possibilities ahead. Doing so is misleading, however, and sooner or later we will have to reckon with a less agreeable reality.

The challenges ahead will be far more difficult than the public has been led to believe and than most of our present leadership apparently understands. Despite the considerable progress in raising awareness of climate change, we are still in a "consensus trance," oblivious to the full scope, scale, severity, and duration of the climate destabilization already under way. Most believe that a few minor adjustments, a few policy changes, and improvements in energy efficiency will be enough to get us through without jeopardizing the "American Dream" or upsetting the consumer society. But a sober reading of the science of climate change indicates something else: we have already set in motion forces and trends that threaten the stability of the biosphere in a few decades and that will persist far longer. Some highly credible scientists like James Lovelock (2009) believe the stability of civilization could similarly fail by the end of the century or even sooner. We are simply unprepared to respond adequately for anything so devastating. If the United States were a sailing ship heading into stormy seas, we would be well advised to lighten the load, secure cargo, trim sails, and batten down the hatches. But no comparable actions are being discussed in the United States or elsewhere. With a few exceptions, climate change is still regarded as a problem to be fixed by small changes, perhaps profitably, and not as a series of dilemmas or as a challenge to consumerism, the growth economy, or—in a more abstract but no less real way—to our institutions, organizations, philosophies, and paradigms.

The crisis ahead is first and foremost a political challenge, not one of economics or technology, as important as those are. The global crisis ahead is a direct result of the largest political failure in history. The U.S. government and elected officials, particularly in recent years:

- Ignored the increasingly urgent and rigorous warnings of danger, and thereby
- Failed to anticipate ecological and climate trends, and so
- Made little or no effort to alert the public to the dangers ahead;
- Were oblivious to the security implications of rapid climate change;
- Took none of the obvious steps to recalibrate the economy to protect natural capital, including climate stability;

- Did little to promote energy efficiency and renewable energy; and thereby
- Wasted trillions of dollars, which helped to weaken the economy and thereby contributed to the collapse of financial markets in late 2008 and
- Created a legacy of debt and deficits both ecological and financial.

Perhaps all of this can be explained by the generally modest level of scientific literacy characteristic of elected officials. Policy failure at this scale certainly reflects the stranglehold of coal and oil money on public policy. And the magnitude of failure has been multiplied by the wasted treasure and time spent chasing the neoconservative mirage of U.S. global domination. Whatever the cause, political leaders of both parties squandered opportunities to act when the crisis could have been headed off for a fraction of what we've paid for the misadventure in Iraq. And decades of such governmental and political failure have brought us uncomfortably close to the brink of global collapse.

The blame cannot be placed solely on government or particular officials, however, for in a democracy government reflects, more or less, the larger public will. Responsibility must be shared by all of us, including notably the media. Long ago Walt Kelly's cartoon character, Pogo, captured this by saying "we have met the enemy and he is us." Climate destabilization, similarly, is the aggregate result of our means of travel, our consumption, the infrastructure by which we are fed and provisioned, and our manner of living, all of which have been subsidized by the rapid drawdown of fossil fuels. The enemy is us... but all of us together, properly led, can make a big difference. And this is where governments enter the picture. The multiple crises ahead require very different public priorities and changes in policy, law, regulation, and the political processes by which we conduct the public business.

There is a considerable movement to green corporations, and that is all to the good. But only governments have the power to set the rules for the economy, enforce the law, levy taxes, ensure the fair distribution of income, protect the poor and future generations, cooperate with other nations, negotiate treaties, defend the public interest, and protect the rights of posterity.[7] Errant governments can wage unnecessary wars, squander the national treasure and reputation, make disastrous environmental choices, and deregulate banks and financial institutions, with catastrophic results. In other words, we will rise or fall by what governments do or fail to do. The long emergency ahead will be the ultimate challenge to our political creativity, acumen, skill, wisdom, and foresight.

It is time for a higher level of realism about our situation and the capacity of people in crisis to respond heroically. Many will disagree. Even at this late hour some are inclined to dismiss out of hand what they call "doom and gloom," preferring to talk about happier things. Believing that people can't handle the truth, they offer instead variations on the theme of "50 easy ways to save the Earth" that threaten neither the lifestyles of consumers nor the power of corporations. Many place their faith in heroic technology of one kind or another. Some believe that climate can be stabilized at a profit, without pain, suffering, or sacrifice. Such views, however, would have been far more plausible 30 years ago. Those casting themselves as "optimists" underestimate the capacity of people to respond while misleading them about the severity of what lies ahead and the adjustments that will have to be made. This book is written in the belief that people want to be told the truth and that with intrepid and competent leadership and encouragement most will rise to meet the realities ahead. And that is the best chance we have to get through the long emergency more or less intact.

I also write with the assumption that we will succeed in reducing atmospheric CO_2 below the level that would cause runaway climate change; otherwise, there is no point in writing anything other than an elegy or funeral dirge. My focus, accordingly, is on the leadership that will be necessary for us to respond politically and morally to what we will have already "bought" up to the point at which the level of all heat-trapping gases in the atmosphere is stabilized and trending downward. The book focuses on three challenges of transformational leadership in the decades and centuries ahead. The first is to prepare the public to understand the scope, scale, and duration of climate destabilization and to grasp the fact that it is first and foremost a challenge to our system of politics and governance. The second is to help us understand the connections between our energy choices and ecological consequences, including those of a deeper sort that we commonly assign to religion. The third is to help forge an honest vision of the future and lay the foundation for authentic hope.

Some believe that we are approaching our "final hour," others that we've arrived at the "singularity," a point at which our minds and bodies will be merged with our machines whether we like it or not. By whatever name, however, we live in paradoxical and perilous times rendered more so by a deficit of vision. If our future were made into a movie and fast-forwarded a few decades, it would have no good ending. But trend is not destiny, as economist Herman Daly pointed out long ago. Destiny is the sum total of the choices we make, and we have the power to make different choices and hence to create a destiny better than that in prospect. The challenge to those intending to lead is to help create a vision of a decent human future within the bounds of ecological possibility. We must honestly face the forces we've set in motion and look to a farther horizon. My subject is hope of the millennial kind.

PART I

POLITICS AND GOVERNANCE

CHAPTER I

Governance

We lack a theory of governance.... We need to invent whole new institutions, new ways of doing business, and new ways of governing.

—Amory Lovins

I favor politics as practical morality, as service to the truth, as essentially human and humanly measured care for our fellow humans.

—Vaclav Havel

THE U.S. CONSTITUTION AND THE BILL OF RIGHTS were drafted in an agrarian era by a small group of men as collectively brilliant as any in history. The government they created was designed with checks and balances and divided authority in order to prevent executive tyranny, sometimes override popular majorities, and avoid quick action on virtually anything. From its agrarian origins it has grown incrementally ever since in response to particular issues, economic necessity, and above all war, but not as a result of much planning, foresight, or effort to create a coherent political architecture.

Nonetheless, the framework they created has survived and even thrived through sectional rivalry and the Civil War, the excesses of the Robber Baron era, two world wars, and the rise and fall of fascism and communism. The Constitution, for some, is a scripture hence beyond reform. Historian Charles Beard, less reverential, once argued that it was written to protect private wealth, especially that of the founders. That may not have been as true as Beard assumed for the founders, but it is clear that "By the middle of the nineteenth century the legal system had been reshaped to the advantage of men of commerce and industry at the expense of farmers, workers, consumers, and other less powerful groups within the society" (Horwitz, 1977, pp. 253–254). More recently, political scientists Robert Dahl, Sanford Levinson, Daniel Lazare, and Larry Sabato have questioned the inclusiveness of the Constitution as well as its effectiveness and future prospects. Dahl, for example, argues that undemocratic features were built into the Constitution because the founders "overestimated the dangers of popular majorities... and underestimated the strength of the developing democratic commitment among Americans" (Dahl, 2002, p. 39; Lazare, 1996, p. 46). While somewhat pessimistic about the prospects for greater democratization, he argues that "it is time—long past time—to invigorate and greatly widen the critical examination of the Constitution and its shortcomings" (pp. 154–156). Constitutional law expert Sanford Levinson agrees: "the Constitution is both insufficiently democratic... *and* sufficiently dysfunctional, in terms of the quality of government that we receive... [that] we should no longer express our blind devotion to it" (Levinson, 2006, p. 9). Accordingly, he proposes a new constitutional convention "to do what the framers of 1787 did," by which he means update and improve the document based on the experience of other democracies and the two centuries and more since the founding (p. 173).[1]

Beyond issues of democracy and inclusiveness are other questions about how well the Constitution works relative to the climate and the environment. The environment is a complex, interactive, and nonlinear system. But the structure of the Constitution favors "decentralized, fragmented, and incremental lawmaking," in legal scholar Richard Lazarus' words (2004, p. 30). As a result, laws, policies, agencies, and whole government departments often work piecemeal and at cross-purposes, without due regard for long-term consequences.² Political scientist Frank Kalinowski argues further that the roots of our environmental problems, such as the rampant individualism that undermines the public interest, the commitment to growth at whatever ecological cost, incremental decision making that blinds policy makers to the connections between air, water, land, wildlife, human health, and long-term prosperity, and the tendency to discount the future, "all can be found fixed in the processes of the Constitution."³ Philosopher Thomas Berry attributes that flaw to the preoccupation of the writers of the Constitution with property rights, "with no recognition of the inherent rights of nature and no defense of the natural world" from corporations (Berry, 2006, pp. 108–109).

Whatever one's views of the Constitution, beginning with the onset of the Cold War government became increasingly shrouded in secrecy and organized to accelerate the exploitation of natural resources, subsidize corporations, treat the symptoms of environmental problems without touching their root causes, alleviate some aspects of poverty without solving deeper problems, and protect the interests of the wealthy. We have had neither an open and honest political system that effectively encouraged public participation in major decisions nor one particularly distinguished by its competence—partly the fault of self-fulfilling prophecies from those who said they wanted to get government off our backs. One predictable result was a marked decline of public confidence in political institutions and widespread cynicism and apathy that undermined democracy and encouraged yet more malfeasance in high places.

These problems were compounded by the response of the Bush administration to the events of September 11, 2001, casting further doubt on the stability and workability of the constitutional arrangements. Specifically, the theory of the "unitary" executive set the precedent for a presidency beyond the reach of Congress and the courts, armed with the power to wage wars and spy on the public with few if any restraints. But long before the misadventures of the George W. Bush administration, the constitutional order was greatly unbalanced by 20th-century wars and global economic forces, much to the advantage of the executive branch.[4]

A great deal of that unbalancing, however, can be explained by our approach to politics and suspicion of democracy in particular and government in general. Historian Garry Wills describes the antigovernment tradition in the United States as the confluence of:

> the lack of a symbolic center (religious or political) at our origins, the air of compromise in our Constitution's formation (which made it vulnerable to the reversal of Federalist and Anti-federalist values), the Jeffersonian suspicion of the Constitution (which Madison abetted at one stage), a jostling of competitive states' claims (reaching a climax in the secession of the South), a frontier tradition, the "Lockean" individualism of our political theory, a fervent cult of the gun. All these were added, in overlapping layers, to the general anti-authoritarian instincts of mankind. (Wills, 1999, p. 318)

The American approach to governance, in Wills' view, leads to the worst outcomes: inefficiency *and* despotism. The antecedents lie in the fact that early settlers came to the New World to escape the overbearing hand of government and to become rich. Americans, consequently, are said to venerate liberty more than anything else, and for many this implies little more than freedom from government. The fear of tyranny fueled the heated debates about the ratification of the Constitution and later about the rights of states to act independently of federal authority, leading to the Civil War. Even in the changing circumstances of industrialization and world wars, many Americans remained suspicious of Washington and centralized authority, but often without the slightest concern about the power of corporations. Our Bill of Rights and political culture celebrated freedom from government, as often noted, but not the kind of positive freedoms that Franklin Roosevelt proposed in his State of the Union address in 1944.[5] The resurgence of conservatism after Barry Goldwater's defeat in 1964 was largely a rebellion against some kinds of government authority but not against the burgeoning militarization of society or the suppression of dissidents or the expanding power of corporations. The result, in political theorist Sheldon Wolin's apt phrase, is a kind of "inverted totalitarianism," representing "the *political* coming of age of corporate power and the *political* demobilization of the citizenry" (Wolin, 2008, p. x). Americans are indeed a people of paradox, confused about the meaning of fundamental terms such as democracy, freedom, equality, liberalism, and conservatism, and about the limits to power on one hand and personal freedoms on the other.

CONVERGING CHALLENGES

As difficult as these issues have been, the hardest tests for our Constitution and democracy are just ahead and have to do with the relationship between governance, politics, and the dramatic changes in Earth systems now under way. Human actions have set in motion a radical disruption of the biophysical systems of the planet that will undermine the human prospect, perhaps for centuries. The crucial issues will be decided by how and how well we conduct the public business in the decades and centuries ahead, but now on a planetary scale. Of the hard realities of governance ahead, five stand out.

The first challenge is that posed by climate change driven by the combustion of fossil fuels and changes in land management. The Fourth Assessment Report from the Intergovernmental Panel on Climate Change (2007), the Stern Review (Stern, 2007), the research on the effects of global change on the United States carried out by the National Science and Technology Council (2008), and other scientific evidence indicate that our future will be characterized by:[6]

- Rising sea levels by perhaps, eventually, as much as five to six meters or more, but no one knows for certain. What is known is that virtually everything frozen on the planet is melting much more rapidly than anyone thought possible even a few years ago.
- Higher temperatures almost everywhere, but concentrated in the northern latitudes, melting permafrost, and boreal forests turning from weak sinks for carbon into sources of carbon and methane.
- More drought and severe heat waves, particularly in mid-continent areas.
- Tropical diseases spreading into regions with previously temperate climates and emergence of new diseases.
- Degradation of forests and ecosystems due to higher temperatures, drought, and changing diseases.
- Rapid decline of marine ecosystems threatened by acidification and higher surface water temperatures.
- Larger (and possibly more frequent) hurricanes, tornadoes, and fires.
- Loss of a significant fraction of biological diversity.

Given our past emission of heat-trapping gases, much of this is simply unavoidable. Regardless of what we do now, the Earth will warm by another half to a full degree centigrade by midcentury, bringing us uncomfortably close to what many scientists believe to be the threshold of disaster. The climate system has roughly a 30-year thermal lag between the release of heat-trapping gases and the climate-driven weather events that we experience. Hurricane Katrina, for example, grew from a Class 1 storm to a Class 5 event quite possibly because of the warming effects of carbon released in the late 1970s.[7] Similarly, the causes behind the weather headlines of the future will likely include the use of fossil fuels and land abuses decades before. We are already committed to a substantial warming of the Earth, by as much as 1.8°C above pre-industrial levels (Lynas, 2007, p. 246).

Many credible scientists believe that we still have time to avert the worst, but not a minute to waste. No one knows for certain what a "safe" threshold of heat-trapping gases in the atmosphere might be. For hundreds of thousands and perhaps millions of years, the level of carbon dioxide did not go above ~280 parts per million (ppm), compared to the present level of 387 ppm, with another ~2+ ppm added each year. Climate scientist James Hansen has recently proposed 350 ppm CO_2 as the upper boundary of safety (Hansen et al., 2008).

We are clearly in uncharted territory. Further delay in stabilizing and reducing levels of CO_2 poses what economist Nicholas Stern calls a "procrastination penalty" that will grow steadily until we eventually cross a point of no return. In other words, it will be far cheaper to act now than at some later date when effective action may no longer be possible. If the warming should occur abruptly "like the ones that are so abundant in the paleoclimate record," we will have no time to adapt before catastrophe strikes.[8] And there is good reason to believe that the climate system is indeed highly sensitive to small changes: "Earth's climate is extremely sensitive: it is capable of taking inputs that seem small to us and transforming them into outputs that seem large" (Broecker and Kunzig, 2008, p. 181).

No matter what our personal preferences, politics, or beliefs may be, as greenhouse gases accumulate in the atmosphere, temperatures will continue to rise until the Earth reaches a new equilibrium. Even were we to stop emission of CO_2 today, sea levels from the thermal expansion of water and increasing mass from the melting glaciers and ice caps would change coast lines for perhaps the next thousand years (Solomon et al., 2009; Archer, 2009). If the rate of melting is rapid or sudden, the migration inland will create hundreds of thousands, or more likely millions, of refugees—like Katrina but on a much larger scale. Unless we choose to build dikes and can afford to do so, many coastal cities will be flooded, possibly within decades or by the end of the century. A majority of the millions of people who live along the Gulf Coast and eastern seaboard will have to move inland to higher ground. But we have neither the money necessary to relocate millions of people nor the infrastructure to accommodate them once moved.

The warming of the northern latitudes and oceans means many things, among which is the possibility of triggering positive feedbacks that will cause the release of large amounts of methane from permafrost and the ocean floor. As with other possible tipping points, a large release of methane to the atmosphere is a wild card in the deck that hopefully will never be brought into play. But again the scientific evidence does not permit us to predict accurately. It is clear, however, that the government is ill prepared to handle the social, economic, and political disruption to which we are now committed, to say nothing of the effects of more rapid changes.

It is especially difficult for Americans to imagine empty supermarket shelves and the possibility of famine. But with each increment of temperature increase, heat waves and drought in the U.S. mid-continent become more likely, jeopardizing much of our food system. A forecast by the Consultative Group on Agricultural Research (news.bbc.co.uk/2/hi/science/nature/6200114. stm; see also Battisti and Naylor, 2009) indicates the likelihood that climate change–driven heat waves, drought, and floods will render much of the Midwest unsuitable for agriculture by 2050. At the very least, recurring droughts and heat waves of longer duration combined with larger and more frequent storms, floods, and changing crop diseases will make farming even more precarious than it already is.[9]

Tropical diseases such as malaria and dengue fever could spread into areas that once had temperate weather. People exposed to excessive heat and higher humidity levels will be vulnerable to entirely new diseases as well. Rapid climate change poses even more severe problems. "The multiple factors that are now destabilizing the global climate system," in Paul Epstein's words, "could cause it to jump abruptly out of its current state. At any time, the world could suddenly become much hotter or even colder. Such a sudden catastrophic change is the ultimate health risk."[10]

Entire ecosystems will be degraded, reducing the services once provided by the particular ensemble of plants and animals adapted to specific places and temperatures. The ecological effects, as complex as they will be, however, are better understood than those that will be imposed on the human psyche. As the once familiar trees, birds, and animals of a region die out, the sense of loss will be impossible to calculate. People, attached to the sights, sounds, and smells of familiar landscapes and regions will go through a process of grieving similar to that of refugees forced to flee their homes and cherished places. The degradation of the forests of Appalachia and the Southeast to scrub and grassland, for example, will incur crushing psychological costs for which we have no adequate words.

The future now on the horizon will be characterized as well by larger and more frequent storms. In coastal areas hurricanes will be more intense, with much larger storm surges spreading devastation farther inland. Rain events will be larger, and the frequency of tornadoes and severe storms will increase. But nature in every part of the Earth will become more capricious and strange. At some point we will indeed have made Earth into a "different planet."

Some may quibble about the timing, but it is clear that we are headed toward a global disaster that has the potential to destroy civilization. But the conversation about changes in governance, economics, social norms, and daily life that must be made to avoid the worst of what lies ahead is only beginning. In short, the level of public awareness and policy discussion does not yet match the gravity of the situation. The prevailing assumption is that we can adopt better technologies like hybrid cars, solar collectors, and compact fluorescent lights and change little else. We will need all the technological ingenuity that we can muster, but the science indicates a much more precarious situation and the need for deeper changes that will require substantial alterations in our manner of living. "There is," in John Sterman's words, "no purely technical solution for climate change…we must now turn our attention to the dynamics of social and political change" (Sterman, 2008, p. 533).

The second challenge, described in *The Millennium Ecosystem Assessment Report* (2005), indicates that our future will likely be full of nasty surprises caused by the breakdown of ecosystems and the ecological services they provide. Changes in land use, encroachment of human populations into formerly wild areas, and pollution, all compounded by rapid climate change, will continue to exacerbate the number and severity of changes amplifying a serious decline in the health of ecosystems, species diversity, and the overall stability of the biosphere. The Earth's systems, including the oceans, are everywhere under assault, with no end in sight. The timing is particularly unfortunate. Ecological degradation radically impairs the resilience of ecosystems to climate change and reduces their capacity to sequester carbon.

A third challenge is that we are approaching the peak of global oil extraction, which could collapse the energy scaffolding that supports modern society, economic growth, and our particular version of the good life.[11] The famous bell-shaped curve of oil extraction developed by petroleum geologist M. King Hubbert in 1956 to portray the peaking of U.S. oil extraction is applicable to the global oil economy as well. The cheapest and most accessible oil has already been exploited. Having exhausted the easiest, cheapest, and nearest sources of oil, what remains is deeper down, farther out, harder to refine, and often located in places where we are not much admired and where the politics are unfathomably contentious. As a result, it has become far more expensive to extract, refine, transport, and defend our access to it. This is not the end of oil but rather the beginning of the end of cheap oil and the way of life we built on the flimsy assumptions of easy mobility, convenience, and dependability of long-distance transport of food and materials. Because we have failed so far to advance energy efficiency and renewable energy as a national policy, when supply and demand curves for oil eventually diverge the results likely will be long gas lines, economic downturn, unemployment, inflation, political instability, and wars fought over the remaining reserves of oil.

The end of the era of cheap oil has been apparent at least since the first oil embargo in 1973, but we failed to take effective action commensurate with the scale of the challenge and with the opportunities created by rapidly improving technology. In November of 1976 I helped organize an effort to inform the newly elected Carter administration about the largest environmental challenge he would likely face. We chose to focus on energy policy, which was then and is now the linchpin connecting the economy, security, equity, and environment. The resulting "Wolfcreek Statement" proposed raising the price of energy by imposing severance taxes on all fossil fuels at the mine mouth or wellhead until the price of fossil energy equaled the marginal cost of the cheapest renewable energy alternative. My coauthors also recommended that the proceeds of the tax be used to help those hardest hit by higher energy prices and to fund research and development on efficiency and renewable energy technology.[12] National politics, however, were dominated by oil and coal companies that blocked all efforts to advance a far-sighted policy. As a result, U.S. dependence on imported oil and fossil fuels grew steadily, even though the technology for fuel efficiency improved dramatically in the same period. The failure of successive presidents and congresses over three decades to create a decent energy policy certainly stands as one of the most egregious policy failures ever, with ramifications that led to terrorism, oil wars, deficits, economic vulnerability, global economic shifts, and climate change. All in all, U.S. energy policy over the past 30 years has been a perfect failure, and since we have had no foresight and precious little leadership, our good options now are fewer than we would otherwise prefer.

The best course ahead, what Richard Heinberg calls "powerdown," requires a rapid shift to energy efficiency, solar energy and other forms of renewable energy, and unavoidable changes in human behavior (Heinberg, 2004). Fuels made from biomass, tar sands, or coal will not have the same energy return on investment and will likely be considerably more expensive and environmentally destructive.[13] With enough foresight, powerdown does not have to be disastrous, but it does mark the end of a century-long energy binge powered by cheap and readily available oil.

The fourth challenge is equally self-inflicted. We have entered a new era in U.S. politics that will be characterized by what political scientist Chalmers Johnson calls "blowback."[14] The U.S. global military presence is maintained by 737 military bases scattered around the world plus an unknown number of secret detention centers, training facilities, and surveillance sites (Johnson, 2006, p. 138). Total military spending, including costs for wars in Iraq and Afghanistan, is estimated to be over $1 trillion annually, well above the official budget of $625 billion (Johnson, 2008). Whatever the real number, our exorbitant military expenditures buy us little safety or security. To the contrary, they ensure economic ruin at home and resentment abroad, raising the likelihood of future attacks on the United States and American citizens. Our adventures in the Middle East will likely trigger terrorist attacks here and elsewhere that have the potential to cause domestic havoc quite independent from that caused by climate-driven weather events or the end of cheap oil.[15] Likely targets include cities, the utility grid, and the Internet. For some defense officials, it is only a matter of when, not whether, such things will occur. The American military presence around the world is the result of many factors, not the least of which is the necessity of maintaining our access to oil in order to perpetuate consumerism a while longer at considerable expense and risk. But we pay for our oil imports mostly by borrowing from lenders, including China. Much of that money goes to people who in turn fund terrorists. All empires, however, eventually rot, subject to arrogance, overreach, debt, and defeat at the hands of resourceful and agile adversaries—and ours is no exception.

The fifth challenge is the necessity to repair the collapsed financial system and reform an economy built on excess, debt, and dishonest bookkeeping. The economic collapse that began in 2008 came as a surprise only to the comfortably drowsy. In the years leading up to the crisis we had ample warnings, including some from insiders like George Soros. But those in charge were paralyzed by the ideology of the free market, the sheer complexity of the global financial and economic system, and the artful practice of greed. Unsurprisingly, they failed to take preventive action in time because they saw nothing to their advantage that needed preventing.[16] The economic and social consequences of that failure will be felt for a long time, and they go well beyond this book. Suffice it here to say that the denominator common to the five is a mind-set that blends the philosophy of "after us the deluge" with the proposition that the devil can have the hindmost. The lesson is that bovine indifference to deficits and debt, whether pertaining to the atmosphere, ecosystems, energy, or even international goodwill, tend to compound each other and sooner or later take their toll on the "leading economic indicators." But by any full reckoning we were never as rich as we thought we were. We were living on credit by drawing down natural capital, as Herman Daly and other ecological economists said long ago.

While each of the five challenges can be described separately, we will experience them as interactive parts of a single long emergency. Each part will amplify the others, generating novel results. The challenge posed by the deterioration of planetary ecosystems will be worsened by higher temperatures, larger storms, and changing rainfall brought on by climate change. The loss of soils and species diversity and the impairment of ecological functions will in various places reduce the capacity of Earth to support life and sequester carbon. And there are thresholds beyond which the capacity of the Earth to support life will be irretrievably mutilated. While the warnings described in the *Millennium Ecosystem Assessment Report* (2005) are no less real than those of impending climate change, the amplifying and interactive effects of ecological decay are harder to describe and dramatize and therefore harder for policy makers and the public to comprehend. Each strand of the long emergency will create conditions in which desperate people may well do desperate things, thereby diverting attention and resources to the headlines of the moment and to relieving symptoms rather than solving underlying causes.

Beyond those mentioned above, other events, trends, and processes will affect the human prospect, notably continued population growth from the present 6.8 billion to upwards of 9 billion, emerging diseases amplified by warming temperatures, and the complexities of global economic and financial interdependence, which are said to be beyond mortal comprehension. The human future, in other words, will be something like a quadratic equation that requires simultaneously solving a series of problems correctly in order to arrive at the overall right answer. The stakes for humankind have never been higher. The challenge is global and beyond the capacity of any one nation to resolve on its own. Our situation does not have to end in catastrophe, but it certainly will unless we act soon to recalibrate economies, political systems, and personal expectations to the realities of the biosphere.

Our capacity to respond to the challenges of the long emergency will be further complicated by a growing backlog of domestic problems. Using data from the Congressional Budget Office, the Center on Budget and Policy Priorities, for example, forecasts a national debt by midcentury of \$40–\$45 trillion, larger than the present world economy (Kogan et al., 2007). That number has been increased dramatically by larger deficits from the 2008 bailout of banks and financial institutions and spending to stimulate the economy in 2009. Decaying infrastructure of roads, water systems, dams, levees, and the utility grid will require trillions of dollars to repair. The U.S. education system continues to turn out a high percentage of young people ill equipped for productive life in a complex society and with little capacity for critical thinking. More than two million Americans are locked away in prisons—a larger fraction of the population than in any other country. On any number of social indicators the United States ranks in the bottom tier of developed nations.

IMPLICATIONS

The implications of the five challenges of the long emergency are becoming clear. The first priority for government now is to take preventive measures to avoid the worst of what could lie ahead (see www.climateactionproject.com and Becker, 2008). This requires urgent steps to reduce our own CO_2 emissions by 90 percent by 2050 and to lead the global effort to prevent a temperature increase above 2°C (Hansen, 2008). An effective climate policy is predicated on an energy policy that rapidly moves us away from fossil fuels before supply interruptions and climate consequences become unmanageable and catastrophic.[17] For those who say we cannot do anything about climate destabilization in the present unstable economic situation, the answer is obvious. Part of the reason we landed in the current economic crisis is that we've been spending upwards of $1 trillion each year on the wrong energy choices and another $600+ billion for imported oil.[18] We have been unnecessarily hemorrhaging a stupendous amount of money for a long time, by one estimate upwards of $48 trillion since 1960. A McKinsey & Company study in 2007, however, showed that we could conservatively eliminate 30 percent of our carbon emissions by 2030 by improving energy efficiency at no net cost (McKinsey & Company, 2007). A more aggressive approach would lead to cuts perhaps as high as 50 percent, still at no net cost. Sharply improved efficiency and desubsidizing coal, oil, natural gas, and nuclear would free revenues that could be put to better use stabilizing the economy and capital markets while building the foundation for a green economy (Jones, 2008). In addition, the auction of permits to release carbon as part of a cap and trade system would generate ~$200+ billion each year, part of which could be used to finance the transition to an efficient solar and wind-powered economy. In short, a major cause of the present economic crisis is energy waste and inefficiency, but by the same logic radically improving energy efficiency and deploying solar and wind technologies can be a major part of the solution because these are the fastest and least costly solutions for multiple problems. Said differently, adoption of a robust energy policy is the fastest and cheapest way to improve the economy, environment, health, and equity and increase security. It is the keystone issue, not just another stone in the arch.[19]

Standing in the way of that transition, however, is an army of lobbyists hired by the coal, oil, and nuclear industries. Among other things, they argue for a "balanced" energy policy, one that "keeps all options on the table." Doing so appears to be reasonable because we have not developed a coherent way to make "apples to apples" comparisons among various alternatives, including efficiency, distributed solar energy, coal, nuclear power, and biofuels. Were we to do so, we would insist that choices be made on the full costs of various options, including:

- The energy required to capture, process, and transport energy in its various forms,
- Opportunity costs measured as carbon removed per dollar invested,
- All environmental impacts, including those of future climate change,

- Costs to human health,
- Costs of all federal, state, and local subsidies, including levies not collected,
- Costs of insurance against potentially catastrophic failure, and
- Social impacts, especially to the poor.

It makes no sense whatsoever to choose policies that switch from potentially catastrophic problems to those that are merely ruinous. Proposals for "clean coal," for example, ought to be evaluated against all of the effects of mining on land, water, and people, as well as the costs and uncertainties of sequestering carbon in perpetuity at a cost that competes fairly with all other alternatives.[20] Similarly, consideration of nuclear power must include the subsidies for fuel enrichment and the costs of insurance against accidents, decommissioning power plants, and storage of high-level wastes in perpetuity, as well as the civil liberty implications of securing the nuclear fuel cycle and guarding its waste products for thousands of years, the effects on weapons proliferation, and an analysis of the risk of catastrophic failure on a Chernobyl scale, whether by acts of God, human error, or malice.[21] In short, compared to every other choice nuclear power is slow, expensive, dangerous, incompatible with democracy, and uncompetitive with benign, cheaper, and more agile alternatives. All energy choices, however, must be measured against the potential for radically increased efficiency at the point of end use, distributed solar technologies, and better design of communities, neighborhoods, and public transport that would eliminate the need for a large fraction of current fossil energy use.

Second, under the multiple stresses described above, it is likely that economic contraction, not expansion, will become the norm. If so, many things that we associate with economic growth will be at risk. Harvard economist Benjamin Friedman, for one, argues that broad-based economic growth is directly related to the moral advancement of society, by which he means greater opportunity, tolerance of diversity, social mobility, commitment to fairness, and democracy (2005, pp. 4–5). The relationship is at least questionable; as per capita wealth has increased, our willingness to help alleviate both domestic and global poverty seems to have declined. It is just as likely that, beyond some threshold, economic growth generates consumerism, selfishness, and egoism, corrodes character, and foreshortens concern (Douthwaite, 1993; Sennett, 1998).

In either case, governments in the long emergency will have to learn to manage the economy under conditions in which quantitative growth will slow and eventually stop. Many otherwise credible analysts, however, recoil at the idea of limits to growth, in part because the ideology of growth has become so deeply embedded in our economic orthodoxy, politics, institutions, and personal expectations that we cannot imagine living with less in a steady-state economy. More seriously, limits to growth would require that we face the daunting political challenge of distributing wealth fairly.[22] Instead, even the most progressive politicians call only for "sustainable development," one suspects without a clue what that means or what it might entail. "What politicians will not admit," in University of Surrey professor Tim Jackson's words, "is that we have no idea if such a radical transformation is even possible, or if so, what it would look like. Where will the investment and resources come from? Where will the wastes and the emissions go? What might it feel like to live in a world with 10 times as much economic activity as we have today?" (2008). What is clear, however, is that growth predicated on the availability of cheap fossil fuels and the belief that we could burn them with impunity is coming to an end. So, too, the power of the developed world to offload the ecological costs, risks, and burdens of economic growth on the third world and future generations. One study shows, for example, that between 1961 and 2000, 87 percent of an estimated $91 billion of global ecological debt was imposed on third-world countries, a number three times their total foreign debt.[23] The doctrine of perpetual growth was also impervious to the evidence of mounting constraints imposed by the global drawdown of natural capital of soils, forests, and resources, and the larger ecological effects of waste disposal on the atmosphere and oceans. The faster economies grew, the greater the cumulative damage. Finally, the idea that the global economy can grow continually in the 21st century requires one to believe that we will make a seamless transition from a profligate consumer-driven economy to an era of natural capitalism, that decision makers will choose wisely, and that corporate chiefs will act for the long-term good—not that there will be what appear later to be mistakes, greed, panic, and a mad scramble to seize whatever one can get while the getting's good. The effect of the boom years was a kind of success trap in which we built economies on the shifting sands of illusion, greed, ill-will, and fear.

Economic growth, as presently conceived, cannot be sustained nor should it be. The economy, in Herman Daly's words, "is now reaching the point where it is outstripping Earth's ability to sustain it" (2008). As Paul Hawken, Amory Lovins, and Hunter Lovins argue in *Natural Capitalism*, there is a better economy to be created that does not depend on drawing down natural capital, imposing costs on the poor or our posterity, confusing prosperity with growth, and risking global catastrophe (1999). But the development of that economy will require clarity about the fair distribution of wealth and risk and shrewd public policies.[24] It will require us to relearn the arts of frugality, sharing, and neighborliness. It will take a bit of ingenuity to craft what Howard and Elisabeth Odum call a "prosperous way down" (2001). But as the largest debtor nation in world history, we have less of a cushion to soften the effects of the downturn than had been presumed. The Congressional Budget Office forecast of the U.S. debt in 2050 cited above does not include the likely costs of climate change and damages from drought, storms, and degradation or loss of ecological services that we take for granted. Nor does it include the costs of possible terrorist events in the United States.

A third implication of the long emergency is that government will be required to take unprecedented measures to relocate people displaced by drought, storms, and continuously rising sea levels. A sizeable fraction of the U.S. population now lives within 100 miles of a coast and is therefore vulnerable to both increased severity of storms and rising sea levels. Hurricanes Katrina (in 2005) and Ike (in 2008) preview what lies ahead as larger and probably more frequent storms batter coasts. The scientific evidence cited above indicates that ice in Greenland and Antarctica is melting much more rapidly than previously thought. As a result, sea levels will eventually inundate coasts worldwide, including U.S. cities such as New Orleans, Miami, Charleston, Washington, Baltimore, New York, and Boston. We may choose to protect some with dikes, but that will be hugely expensive and probably doomed to failure if global temperatures increase much beyond 2°C. In that situation, the more likely scenario is that millions of people will be forced to leave their homes and property and move to higher ground. But it is not just the people living along coastlines who are at risk. As the mid-continent becomes hotter and drier and subject to more severe tornadoes, storms, and floods like that of 2008 in Iowa or worse, the region will become less habitable as well. As rainfall diminishes in the Southwest, one plausible scenario is that:

> Businesses and families begin to abandon Phoenix, creating a *Grapes of Wrath*–like exodus in reverse. Long lines of vehicles clog the freeways, heading east toward the Mississippi and north toward Oregon and Washington. Burning hot, parched, and broke, the city that rose from the ashes achieves its apogee and falls back toward the fire. (Powell, 2008, p. 240)

In short, governments will have to relocate and house growing numbers of people. In the decades ahead, we must prepare for a future in which large storms, flood, fire, and drought, as well as acts of terrorism, will become the norm. The capacity of emergency management will have to be made much more robust and effective, not just for intermittent events but for multiple events, which may occur regularly. When climate change–driven emergencies become normal, government must have the capacity to quickly and effectively rebuild shattered communities and economies on a more resilient basis.

A fourth implication follows. In the foreseeable future the food system will become increasingly vulnerable to higher prices for fertilizer, pesticides, and fuels for farm operations, transport, processing, and distribution. Moreover, the stresses of higher temperatures, prolonged drought, changing crop diseases, and storms associated with climate destabilization could reduce farm output. The study cited above (p. 20) by the Consultative Group on International Agricultural Research indicates that by 2050 much of the American Midwest will be unsuitable to grow wheat.[25] In other words, after the peak of oil extraction and in a greenhouse world, the dependability of food supply cannot be taken for granted. To till, harvest, process, transport, and market food 1,500 miles from farm to kitchen, the present agricultural system is said to require a dozen fossil fuel calories for each food calorie. All of this is to say that in the long emergency ahead we may plausibly expect that governments once again will have to deal with the ancient scourge of famine, new technologies notwithstanding.

A final implication of the long emergency concerns the necessity of mobilizing a coalition large enough and steady enough to change our politics, economy, and manner of living to fit biophysical realities. The problem is that many people tend to deny bad news, especially when it is complicated and solutions may be costly and inconvenient. Improving the state of the environment has long been that kind of problem. From Plato's observations about soil erosion in the hills of Greece in the 4th century B.C. to George Perkins Marsh's observation in 1864 that humans were everywhere a disturbing environmental force, no government and no society took the evidence seriously enough to do much about it. The reasons are not hard to find. The rate of environmental change was often slow enough as to be virtually invisible to any one generation, which assumed the deterioration to be natural—the problem of "shifting baselines." Many of the causes of ecological deterioration were simply unknown or operated at a scale—either too large or too small—or at a velocity—either too fast or too slow—that we could not comprehend. And much enamored of science and technology and blinded by the 18th-century ideology of progress and economic growth, we did not see what was happening right before our eyes. In the 20th century, the warnings came more often and helped to launch the environmental movement of the 1960s. But the totality of the problem remained shrouded in political controversy and concealed by complexity and was mostly ignored by much of the public, which was happily diverted by consumption and mass entertainment. In the opening decades of the 21st century, the evidence of global environmental collapse is unmistakable—the result of decades of failed policies and inadequate remedies (Speth, 2008).

But how do we say such things to the public effectively enough to galvanize a constituency for significant changes in government, business, and daily life? On one side of the debate are many with a deep concern about the state of the environment who believe that the public, told the full truth, would either ignore it or become despondent. On the other side are those who believe that the only chance to mobilize the public in the brief time available is to speak the truth without exaggeration, but without diluting it with "happy talk." There is, in other words, a serious difference between those who, like T. S. Eliot, believe that "human kind/ Cannot bear very much reality" and those who advocate an approach like that of Winston Churchill in 1940, who summoned the British people to a considerable level of heroism, while offering only "blood, toil, tears, and sweat." With bombs falling on London, Churchill did not talk cheerily about the new opportunities for urban renewal or the possibilities for defeating the Nazis at a profit. But one of the unknowns of our time is whether we are still the kind of people who can be summoned to heroism when it's all on the line. I write in the belief that we are, but more than ever before a great deal of that optimism is based on the hope for wise and farsighted leadership at all levels.

Our models of great leaders, however, are most often military figures in situations in which the stakes were clear, the adversary dependably loathsome, and the duration of the crisis reasonably short. Public morale was expressed as fierce devotion to the nation or the cause until a final victory quickly won. Morale in the century or more ahead, however, will require an extraordinary stamina and more extensive loyalties that unleash creativity, not animosity. In such circumstances, people will respond with greater intelligence and alacrity if they see themselves as part of something noble, not just as consumers and cogs in an economic machine. In short, beyond better technologies and policies, morale in the years ahead will depend on a widely shared vision of a livable future in radically altered conditions.

The list of things to be done is long. We must, accordingly, summon the clarity of mind necessary to separate the urgent from the merely important and identify strategic leverage points where small changes will generate large effects. To stabilize and then reduce concentrations of greenhouse gases we must make a rapid transition to a resilient economy organized around energy efficiency and solar energy while reversing ecological deterioration and fending off probable terrorist attacks on cities and critical infrastructure. On the global level the United States must help to lead the effort to forge a global bargain that fairly distributes the costs, risks, and benefits of the transition within and between generations. Theologian Thomas Berry calls this our "Great Work" (Berry, 1999).

GOVERNMENT AND MARKETS

But how will we organize to accomplish that Great Work? We presently have no system of governance adequate to the stresses and challenges of the century ahead. This fact has led many to believe that we must put our faith in the corporations as the primary agent of change. Indeed, for the past 30 years we have been exposed to a long and increasingly tedious celebration of markets and an equally vigorous denigration of government.[26] Much of this was self-serving hype by people with much to gain from less regulation, lower taxes, and little public scrutiny. Much of it, too, was driven by the ideology of market fundamentalism that thrived in some university economic departments "through a remarkable level of conformism" and throughout the extreme right of American politics through the mystical power of true belief (Saul, 2005, p. 33). Some of it emanated from right-wing think tanks created for the purpose of spreading the worldview once held by the robber barons with all of the compassion of the Social Darwinists and Calvin Coolidge's sense of public urgency. The defense of free markets, in particular, was exaggerated and misleading, and some of it destructive and dangerous.[27]

Abstractions such as the corporations and the market have no interest in the long-term collective future beyond pecuniary gain. Corporations are bundles of capital dedicated to near-term profitability of stockholders, not to the long-term sustainability of the human enterprise.[28] Allegedly given the rights of persons in 1886 in an otherwise insignificant case about payment of back taxes, corporations have steadily acquired sufficient political clout to prevent changes in law and regulation that would infringe on profits.[29] The power of K Street lobbyists in Washington did not change when Tom Delay departed Congress. Money, access, and power are as seductive to Democrats as to Republicans. The implosion of Enron and more recently the collapse of the housing and financial markets are instructive not as aberrations but as a persistent tendency in a system that has been constructed around rules that protect the rights of capital without the countervailing power of an alert government willing and able to protect the public interest. Jonathon Porritt, the chairman of the U.K. Sustainable Development Commission, puts it this way:

> it is clearly politicians who have to make the most decisive inter-
> ventions... it's governments that frame the legal and constitutional
> boundaries within which individual citizens and corporate enti-
> ties must operate; it's governments that set the macro-economic
> framework through the use of fiscal and economic instruments;
> and it's governments (by and large) that set the tone for public
> debate and that can take the lead on controversial and potentially
> divisive issues."[30]

In other words, "the *current* approach to corporate responsibility simply isn't up to the task in hand.... the *primary* responsibility for making it all happen still lies with government" (p. 220).

To work effectively, in other words, markets have always required energetic, flexible, and imaginative governments to set the rules, level the playing field, enforce the law, and protect the larger public good over the long term. That is only to say, as Yale political scientist Charles Lindblom argues, that "the market system can be understood only as a great and all-pervasive part of the structure and life of society," not the other way around (Lindblom, 2001, p. 277). Or, as Amory Lovins puts it, "markets are only tools. They make a good servant but a bad master and a worse religion... That theology [economic fundamentalism] treats living things as dead, nature as a nuisance, several billion years' design experience as casually discardable, and the future as worthless" (Hawken, Lovins, Lovins, 1999, p. 261).

Corporations can do a great many things better than they have, and markets can be harnessed to better purposes than they have served in the past. Some major corporations such as Wal-Mart are greening their operations and supply chains. Others have joined together in the U.S. Climate Action Partnership to support climate legislation. Some survivors in the financial man-agement community are developing instruments and investment tools to shift assets to long-term value aligned with ecological health. But many, in Peter Senge's words, "fail to summon the imagination and courage to face the fact that they are selling the wrong products...to the wrong customers" (Senge, 2008, p. 310). Few will, without strong, imaginative, and farsighted government leadership of the kind we associate with the founding of the United States, Lincoln's response to the secession of the Southern states, and Franklin Roosevelt's leadership in the 1930s and during World War II. Corporations acting in disorganized or unregulated markets will not act consistently for the public good when it no longer serves their short-term shareholder interests. To do otherwise would be fatal to the management of underperforming companies. The cardinal rule of capitalism is to make money, and no amount of greenwashing can hide that fact.

A great deal, accordingly, depends on how and how well we repair and enhance the capacity of government to do what only governments can do. The market is the arena in which we say "I" and "mine" and in which we act mostly for near-term advantage. Government is one in which we come together to say "we" and "ours," in order to protect and enhance our common interests immediately and over the long term. Markets seldom act for the enduring public good; governments can and must. But a great deal of our commonwealth, common property, and capacity to act collectively has been squandered in the past four decades, diminishing our democratic heritage and reducing our capacity to respond collectively to the kinds of emergencies that will become more common in the future. The miserable performance of the Federal Emergency Management Agency in the wake of Katrina and more recently the total failure of regulation that culminated in the bankruptcies of major financial institutions, for example, were the predictable results of decisions by people who wanted to get government off their backs.[31] Both cases (as well as others) starkly revealed a void where we need the capacity for foresight, competent actions in emergency situations, transparency, and accountability. Once again, we painfully learn that assigning foxes to guard the henhouse is good neither for the hens nor eventually for starving foxes.

This is neither an argument against markets in their proper place nor one against corporations properly chartered and regulated for the public good. It is decidedly not an argument against private enterprise, although we have every reason to dislike unaccountable corporate power, as well as the power of business to manipulate appearances so as to appear considerably better than they are. My position is not "socialist," whatever that word is presumed to mean, but it is decidedly in favor of placing limits on corporate power and even individualism where its excesses cast long shadows on the prospects of our grandchildren and theirs. It is not a hymn to a mythical American past but a call to draw strength and perspective from our history, which at its best has been always pragmatic and experimental. We must repair and enhance our civic culture and our collective capacity to solve problems associated with climate change in the brief time before they become unmanageable. To that end we will need courageous leadership and a media sufficiently committed to the larger public good to promote a national conversation on the rules and procedures by which we make crucial choices in the long emergency ahead, starting with those set down in our founding as a nation. In this conversation, business and commerce clearly have an important role. But they can no longer be given the power, whether by domination of the media or by backroom lobbying, to conflate corporate profit with the public interest.

GOVERNANCE AND PUBLIC ORDER
IN THE LONG EMERGENCY

No one can know the founders' "original intentions" on any number of issues, or what they might have thought about those of our time. But we know that they would not have wanted us to end the bold experiment they launched in self-governance or jeopardize the rights and liberties of posterity. And they certainly would not have wished us to risk the future of life on Earth for any reason at all. But they gave us no formula for governance, only the example of an intrepid, ingenious, pragmatic, and partial experiment in democracy.

The founders responded courageously and brilliantly to the challenges of their time, but those pale beside what we can anticipate in the century ahead. They could presume a stable climate and the resources of a mostly untouched continent to meet the needs of a vastly smaller population that lived predominantly on current sunlight, albeit often carelessly. We, by contrast, are a population of more than 300 million and will grow perhaps by another 100 million before our population peaks. We are an island of affluence in a world of 6.8 billion that will peak, perhaps, at 9 billion. We live on the remainders of once vast natural stores of minerals, soils, and forests. We are powered mostly by ancient sunlight in the form of coal and imported oil. We have technology that the founders could not imagine, but that prowess carries risks that would have given them reasons to act more cautiously than we do. All of this is to say that the challenges ahead differ in scale, complexity, velocity, and duration from any we have faced before. Our response, accordingly, must be at least as ingenious, wise, and adaptable as theirs was to the challenges of creating a republic in their time.

Like the founding generation, we need a substantial rethinking and reordering of systems of governance that increase public engagement and create the capacities for foresight to avoid future crises and rapid response to deal with those that are unavoidable (Grant, 2006, pp. 221–237). In the duress ahead, accountability, coordination, fairness, and transparency will be more important than ever. We will have no slack left for corruption, cronyism, secrecy, and incompetence. We will need governments at all levels, as Peter Senge says of business, with "a more robust organizational ecology...that is in tune with the larger living world and more capable of confronting the host of Industrial Age imbalances threatening our biosphere and our societies" (Senge, 2008, p. 356). We founder, however, in the effort to reform governments, mostly because of the power of vested interests and the lack of a sense of urgency. As a result, dozens of blue-ribbon commissions over many decades have made recommendations to improve the performance of various aspects of the federal government, to little lasting effect. They mostly gather dust on the shelves of the Library of Congress.

Even with a more rational and better-informed citizenry and improved means by which its will is expressed, is it possible to improve the performance of government? For many raised on the ideas of Ayn Rand and Milton Friedman, and inclined to believe campaign slogans promising to lower taxes and get government off our backs, the answer is no, mostly because of the alleged incompetence of one government agency or another or because of stories about the behavior of an obscure overbearing bureaucrat. Some of this is pure fantasy and some is the result of self-fulfilling prophecies, but most of it is, in economist Eban Goodstein's words, "a self-consciously manufactured, anti-government ideology that has paralyzed our nation" (Goodstein, 2007, p. 141). Over the last two decades, the upshot is that some agencies and functions of government, like the congressional Office of Technology Assessment, were abolished. Others deemed inconvenient but politically popular were put on starvation rations and staffed with people who did not believe in government. But other parts, notably the military and surveillance functions, were force-fed. Unsurprisingly, with less money and leadership, morale in many agencies plummeted and much of the federal government performed dismally as predicted, justifying still more budget cutting. As a result, the present administration faces a long rebuilding effort to restore morale, competence, professionalism, and purpose to many federal departments and agencies.

It is possible to undo the damage of decades of neglect and to equip government to meet the conditions of the long emergency. But to do so will require creating the capacity necessary to solve multiple problems that cross the usual lines of authority, departments, and agencies as well as those between federal, state, and local governments. In the long emergency, governments at all levels will have to be smarter, more farsighted, more agile, and more strategic. That does not necessarily mean a larger and more intrusive role, but rather one that steers more effectively by incremental adjustments and not by revolution.[32] We will need to build new alliances between the public, nongovernmental organizations, local and state governments, and business. Above all, government must enable creative leadership at all levels of society, and it must lead first by example, not simply by fiat. It must help catalyze the redesign of infrastructure, food systems, communities, transportation, and energy systems that are resilient and secure by design. Every increase in local capacity to grow food, generate energy, repair, build, and finance will strengthen the capacity to withstand disturbances of all kinds. Distributed energy in the form of widely disbursed solar and wind technology, for example, buffers communities from supply interruptions, failure of the electrical grid, and price shocks. Similarly, a regionally based, solar-powered food system would restore small farms, preserve soil, create local employment, rebuild stable economies, and provide better food while reducing carbon emissions and dependence on long-distance transport from distant suppliers.[33] The primary goal in rethinking development and economic growth is to create resilience—the capacity to withstand the disturbances that will become more frequent and severe in the decades ahead.[34]

BOX I.I. GOVERNING LAND USE IN SHIFTING
CLIMATES

The last time we in the United States tried to do anything at the
national level about land use policy was in 1973. That limited effort
was a bill (S.268) introduced in the U.S. Senate by Henry Jackson
that aimed only to provide funds for those states bold enough
to engage in land planning. Toothless though it was, the bill was
defeated with much patriotic chest thumping. And the Republic still stands, or more properly sprawls, having reportedly lost
an average of one million acres to badly planned "development"
each year ever since and another million or so to soil erosion.
In truth, we barely keep track of such numbers, preferring
to take comfort in the total land reservoir of 2.2 billion acres
that has so far buffered us from the consequences of bad judgment and the absence of intelligent planning. But the true costs
of land lost to development and agricultural mismanagement are
considerably larger even than the little that we do count. First,
sprawling development requires more roads, wires, pipes, con-
crete, and materials than does more condensed development or
"planned unit development." A 1974 report by the president's
Council on Environmental Quality concluded that "planned
development of all densities is less costly to create and operate
than sprawl" (Council on Environmental Quality, 1974, p. 7).
Second, sprawl requires lots more energy to move more people
and goods longer distances, and thereby commits this and other
land-use intensive nations to use more oil than they otherwise
would need, leading to foreign policies predicated on dependence
that lead in turn to belligerence or begging. Third, sprawl was
often financed on a foundation of sand now washing away in
a tsunami of bad debt and insolvency. Fourth, sprawl is bad for
our health. Children have to be hauled to soccer practice or to
school, thereby beginning a vicious cycle that leads to obesity
and future health costs for Type II diabetes, heart disease, and less
familiar ailments (Frumkin, Frank, and Jackson, 2004). Sprawl
tends to disconnect children from nature, causing what Richard
Louv calls "nature deficit disorder" and mental problems that arise

(continued)

BOX I.I. (*continued*)

from the lack of healthy contact with living things (2005). We know as well that sprawl destroys natural habitats and is a main driver of the loss of species. Sometimes smarter development can lessen impacts on wild habitats, but the aggregate effect of any new development is probably never positive. And finally, sprawl contributes to the use of fossil fuels and to the loss of carbon sinks (including forests and soils) that are driving climate change.

The news about land in the Fourth Assessment Report from the Intergovernmental Panel on Climate Change is grim, as noted above. Assuming that we are able to cap the warming below a 2.0°C increase, land-use changes, nonetheless, will be dramatic if still somewhat conjectural. Sea levels will continue to rise, perhaps for another 1,000 years, inundating coastal regions. Larger storms will batter coasts, and bigger storm surges will reach farther inland. Mid-continental areas will likely become hotter and dryer, possibly leading to the abandonment of millions of acres that were once breadbaskets. Rainfall events will become larger, with more floods like those in Iowa in June 2008. More frequent tornadoes will stress our emergency response and rebuilding capabilities. Some inland lakes will lose much of their present volume, radically altering shorelines. Lake Erie, for one, is projected to lose 40 percent of its present volume by 2050. Forested regions will be degraded by larger and hotter fires until little is left to burn.

John Locke and others from whom we derive our foundational ideas about land law reckoned with none of this. For Locke, land became private property once someone in the distant past mixed their labor with the land. More than three centuries after Locke, defenders of private property such as legal scholar Richard Epstein propose that property rights ought to be essentially inviolable. The right of governments, then, to take privately held property ought to be confined to a small number of instances in which the taking redounds to the larger good, not just to a larger government (Epstein, 1985 and 2008). The upshot for Epstein and others of his persuasion is that the property rights of farmers, developers, private landowners, and corporations engaged in mining, logging,and energy extraction ought to be beyond the reach of government except in the most extreme cases of public need. Epstein's objections notwithstanding, the law has in fact been excessively kind to the rights of individual and corporate owners of land under the presumption that seizure of privately held land for public purposes ought to be compensated as an otherwise unwarranted taking proscribed by the terms of the 5th and 14th amendments to the U.S. Constitution. But the institution of private property, despite its many virtues, has often sacrificed community goods under the guise of protecting freedom (Freyfogle, 2003).

Property law and land policy built over the past three centuries presumed that climate would be more or less stable and that climate was God's business anyway, not ours. Human-driven climate destabilization, however, will dramatically challenge our views of land, private ownership, and public necessity. Global warming will lead to the inundation of coastal areas and larger and more frequent storms. These will create demands for expensive remedies, including massive earthworks built on land taken from private owners and funded by raising taxes. But at any more than a one-meter rise in sea level, millions of people will have to be moved inland here and elsewhere, and flooded property along low-lying coastal regions will be worthless. So, too, will land in mid-continent areas that will likely dry out under prolonged drought and heat. It is difficult to imagine where climate refugees go to find relief or whose property is to be taken to provide land for housing and new infrastructure. Complicated and bitter disputes will attend proposals to transfer water from, say, the Great Lakes to the Southwest or Far West suffering permanent drought. Liability issues pertaining to the mounting damages from climate change will grow increasingly contentious, rather like the tobacco lawsuits only more so. Like the tobacco companies, no company engaged in extraction and sale of coal, oil, or natural gas can say that they did not know the consequences of what they were doing.

(continued)

BOX 1.1. *(continued)*

John Locke's view of property rights has been particularly influential in the development of property law, but there is another and less appreciated aspect of Locke in which he argued that: "For this *Labour* being the unquestionable Property of the Labourer, no Man but he can have a right to what that is once joined to, at least where there is *enough, and as good left in common for others*" (emphasis added; Locke, 1965, p. 329). Men were entitled only to "As much as any one can make use of to any advantage of life before it spoils; so much he may by his labour fix a Property in. Whatever is beyond this, is more than his share, and belongs to others. Nothing was made by God for Man to spoil or destroy" (p. 332).

In a mostly empty world such caveats were conveniently overlooked. But in a "full world" they will become more important, and they raise many complexities. For example, ownership of land, whether by corporation or individual, is singular, but "as much and as good" applies less clearly to any single entity, hinting at something more like collective rights of a community or even later generations that Locke did not discuss.

What does it mean, for example, for one generation to leave as much and as good for later generations? What might that standard imply for land law largely built on the rights of the living? Application of that standard leads to consideration of how to preserve land and its health for subsequent users and of the conditions that affect land, such as temperature and rainfall, presumed by Locke to be outside our control and responsibility. It is not difficult to extend the argument to include limits on activities that violate the standard of "as much and as good" more broadly to those factors that threaten subsequent generations' access to food, water, and security against storms magnified by the climate-forcing actions of earlier generations.

This leads to a broader interpretation of "takings" applicable to cases in which future generations could be deprived of life, liberty, and property without due process of law. The law as presently interpreted provides grounds neither for solace nor recourse against intergenerational takings, partly because of the complexity of assigning liability, establishing harm, and adjudicating the interests of the parties, one of whom does not exist and the other being too diffuse to name. But such perplexities do not diminish the reality of the deprivation.

If one accepts the possibility of intergenerational takings *and* the limits of remedy available in the present law, the proper course of action is to be found in the arcane and much depreciated activity called planning and in its enactment as effective policy. In plain language, we—the present generation—would have to decide what is properly ours, and further decide not to transgress that line. We would have to further decide the policy means by which to enact those restrictions on all levels of land ownership. In economist James Galbraith's words, planning to prevent the worst of climate change will "empower the scientific and educational estate and the government... it must involve a mobilization of the community at large, and... will impose standards of conduct and behavior and performance on large corporate enterprises" (2008, p. 175).

The idea of national planning is not as far-fetched as it might first appear. We developed comprehensive national plans to mobilize and fight two world wars. Now we face larger challenges. Climate change, the end of the era of cheap fossil fuels, population growth, and ecological degradation are converging to form a global megacrisis for which there is no precedent. But the present policy and legal apparatus for managing land, air, water, energy, and atmosphere in the United States and globally is fragmented, incremental, reactive, and short-sighted. It is imperative that we extend policy and legal horizons to deal with larger systems over longer time periods, much as envisioned in 1969 in the National Environmental Policy Act (NEPA), which requires federal agencies engaged in activities that have the potential to significantly harm the environment to assess environmental impacts, including potential harm to later generations, and identify "irreversible and irretrievable commitments." NEPA was a step toward the kind of integrated and systemic policy planning that we urgently need, but to our great detriment it has been largely relegated to obscurity and ineffectiveness. The principles of NEPA ought to be dusted off, updated with current scientific knowledge, and serve as the basis for reconsidering land use law, beginning with the management of the roughly 700,000,000 acres of farm, rangeland,

(continued)

Box 1.1. (*continued*)

and forest lands. And the three essays that follow below sketch the case for extending our planning and policy horizons out 50 years or more in each of these areas. As the grip of climate change tightens, however, we may discover that present law is inadequate to protect either the present or future generations. It may be that the entire system of ownership will have to be extensively modified in favor of what Peter Brown calls "the trust conception of government," which draws much from Locke's "as good and as much" standard (1994, p. 71). Brown and others, including legal scholar Eric Freyfogle, propose that land law be broadened to include the wider community of life and extended in time to the include rights of future generations. In important respects this is a return to the ancient traditions of English law embodied in the Magna Carta, which included two charters. The first concerned the political and juridical rights of the nobles; the second, and lesser known, was called the Charter of the Forest and guaranteed the rights of people to use the forest and all of its resources as common property (Linebaugh, 2008). It was an economic document that rested on the obvious fact that political and legal rights are meaningless unless undergirded by guarantees to food, water, and materials.

The English commons was eventually whittled down by the conversion of common lands into private property, a process known in history as enclosure. In our time the age-old struggle between enclosure and public access to the commons continues, but at a global scale. The battle is now being fought over control of the common heritage of humankind, including forests, freshwater, the oceans, minerals, genetic resources, the atmosphere, and climate stability. In each case, the powers of exploitation propose to fragment whole systems into pieces, extend the rights of private ownership over common property resources, preserve the domination of a single generation over all those to come, and shorten our policy attention to a few years. The challenge, as noted by poet Gary Snyder, is to create the policy and legal basis that works "on a really long time frame"—"even a few centuries may be insufficient"—so that there will be as much and as good for others (2007, p. 40)

CHAPTER 2

Late-Night Thoughts about Democracy in the Long Emergency

We can have democracy in this country, or we can have great wealth concentrated in the hands of a few, but we can't have both.

—Justice Louis Brandeis

There is only one way to strive for decency, reason, responsibility, sincerity, civility, and tolerance, and that is decently, reasonably, responsibly, sincerely, civilly, and tolerantly.

—Vaclav Havel

DEMOCRACY, WINSTON CHURCHILL ONCE FAMOUSLY said, is the worst form of government except for all of the others ever tried. The Greeks, from whom we inherited the idea of self-government, after all, couldn't manage it for long and fell victim to the political vices of greed, hubris, imperial over-reach, and ruinous wars. In modern times it is possible, historian Walter Prescott Webb once wrote, that the upsurge of democracy in the early modern era was largely the result of the abundance of resources resulting from the discovery of the New World rather than from any general human improvement.[1] His point was that the larger per capita ratios of land, minerals, and natural resources after 1492 reduced the pressures on governments and populations under conditions of scarcity and otherwise diverted peoples' energies to the tasks of getting rich and getting on in the New World, the effect of which was to make us a more agreeable and more manageable lot. The ratios of resources to people, however, are now about what they were prior to the "discovery" of the New World, and the due bill for the long binge of fossil fuel–powered modernization is said to be in the mail. In a more crowded and hotter world, perhaps democracy will be "just a moment in history," as Robert Kaplan (1997) once put it, a casualty of the failure to manage growing complexity and scarcity.

Many other forces also work against democracy. Vice President Al Gore, for one, argues that decades of television and nonstop exposure to advertising have eroded our capacity for the reasoned judgment necessary for democracy and that this is a large factor in the tide of irrationality that has recently flooded our politics. Susan Jacoby, similarly, believes that we live in a "new age of unreason," that America is "ill with a powerful mutant strain of intertwined ignorance, anti-rationalism, and anti-intellectualism," and that Americans are "living through an overarching crisis of memory and knowledge involving everything about the way we learn and think" (2008, pp. xx, 309). The evidence is all around us. Americans watch an average of more than four hours of television each day and are well versed in the lives and doings of celebrities, but many are utterly mystified about politics, history, world affairs, and geography, among other things. Over a lifetime they are marinated in several million advertisements (Barnes, 2006, p. 122) aimed to keep them in a perpetual state of infantile self-gratification as dependable and dependent consumers rather than as informed, active, engaged, and thoughtful citizens. This is not just happenstance but the predictable result of a technological revolution of television, computer games, cell phones, iPods, and sophisticated methods of marketing. Early pioneers in the craft of demand creation in order to sell more stuff than anyone really needed, notably Sigmund Freud's nephew Edward Bernays, believed that the appeal to our lesser side was essential both to the art of selling and to the "engineering of consent" that he considered necessary to maintain the stability of democracy in mass societies.[2] People distracted by consumption, which is to say the thoroughly infantilized, seldom disturb the public order or become zealous revolutionaries. Bernays' methods—the appeal to fear and resentment above all—have been adapted to U.S. politics with a degree of artfulness that might have surprised even Bernays.[3] But it would not have astonished the many critics of democracy from ancient Athens to the present who believe that people in general are ignorant, foolish, gullible, selfish, and incapable of sustained rationality. Harvard economist Joseph Schumpeter, for example, believed that "the typical citizen drops down to a lower level of mental performance as soon as he enters the political field" (1962, p. 262). The American voter, according to Rick Shenkman, is pretty stupid—distracted by consumption, ill informed, lazy, and mentally deficient—and ripe for the plucking, but he proposes only more civic education (2008, pp. 177–179). Other observers blame the lack of leadership. Classical scholar Loren Samons, for example, asks: "When was the last time modern Americans heard a politician, journalistic commentator, or even a character in a popular film openly claim that what the majority—what 'the American people'—want or think is either morally wrong or intellectually bankrupt? But that's what real leaders must do, *especially* in a democracy" (2004, p. 201).

Whatever the various causes, the skeptics of democracy seem to be vindicated by our political life in recent decades. The depth of our political discourse has been mostly inversely proportional to the gravity of the issues. Abortion, for one, has generated great controversy and media attention, while issues having to do merely with the mutilation of life on the Earth, which is to say the abortion of the human prospect, have been mostly greeted by something ranging from awkward silence to ridicule. Sigmund Freud and Carl Jung thought that our political behavior reflected the deeper storms and currents of the subconscious. And after the wars, the gulags, and the many killing fields of the 20th century, who can say what demons lurk below the surface of the conscious mind waiting to march en masse to serve the most heinous causes? But as James Madison noted long ago, the need for government originates not in our virtues but in our frailties, faults, and failings. If men were angels, as he put it, no government would be necessary. In his later years Madison, who more than anyone else authored the Constitution and the Bill of Rights, was pessimistic about the future of the American experiment in democracy. He believed that under the pressures of sheer geographic size and population growth it might have a century "before the rot set in" and democracy came undone (Matthews, 1995, p. 212). But what are the alternatives to democracy?

In 1968, biologist Garrett Hardin published probably the most famous essay ever to appear in *Science* magazine, with the memorable title "The Tragedy of the Commons" (Hardin, 1968). Overpopulation and global environmental problems, Hardin argued, were analogous to the abuse of the common grazing areas of medieval England. As long as the costs of abusing the commons were shared by all and the gains captured by the abuser, the temptation for any one actor to overgraze and eventually ruin the commons for all was overwhelming. "Freedom in a commons," Hardin wrote, "brings ruin to all." The only way to escape tragedy was to increase the collective power to control everyone in the commons, or as he put it "mutual coercion, mutually agreed upon." Economist Robert Heilbroner similarly concluded that mounting ecological threats to human survival could be managed only by authoritarian governments, saying "I not only predict but I prescribe a centralization of power as the only means by which our threatened and dangerous civilization will make way for its successor" (Heilbroner, 1980, p.175). Political scientist William Ophuls concurred, proposing that "ecological scarcity in particular seems to engender overwhelming pressures toward political systems that are frankly authoritarian.... Leviathan may be mitigated but not evaded" (Ophuls, 1992, p. 216). Heilbroner and Ophuls believed that survival in times of ecological scarcity would require a great deal of sacrifice and a lot less consumption, and doubted that the public could, on its own, discipline its appetites sufficiently to avoid disaster. James Lovelock agrees, saying, "We may need restrictions, rationing and the call to service that were familiar in wartime and in addition suffer for a while a loss of freedom. We will need a small permanent group of strategists who, as in wartime, will try to out-think our Earthly enemy and be ready for the surprises bound to come" (2006, p. 153).

There are good reasons, however, why the case for authoritarianism, in particular that made by Hardin, is not wholly convincing. The history of common property resources, for example, reveals that often they were well managed for centuries until outside forces upset the balance of cultural and social restraints on individual behavior. In the English case from which he drew the analogy, the common grazing areas disappeared not due to mismanagement but because they were seized by aristocrats over several centuries of enclosure. Their aim was a more "efficient" and profitable agriculture, and they were willing to displace tens of thousands of smallholders and subsistence farmers whose suffering was justified on the grounds that it supposedly served the larger good. On the scale of history, our experience with coercive governments in general is dismal. The response of the Soviet Union to environmental deterioration, for example, ranged from awful to abysmal. In that and other cases, the reasons include the inflexibility of bureaucracies, the lack of agility in changing circumstances, obliviousness to societal needs, and the many pathologies of unaccountable power that tend to corrupt absolutely (Orr and Hill, 1978).

Still, Hardin, Heilbroner, and Ophuls have a point that has become more urgent with the passing decades. Even if we hold CO_2 levels below the threshold of runaway change, we have already committed to a future with unprecedented ecological stresses. The ripple effects will include economic disruption, political unrest, increasing turmoil in financial markets, growing numbers of climate refugees, wars fought over diminishing resources, and international turmoil. Some in the global economy will make substantial profits in those conditions, but most will not. We may plausibly expect that public order here and elsewhere will be severely tested, and in a growing number of cases it will fail, as it has in Darfur and the Dominican Republic. The conditions ahead could be analogous to wartime emergencies that require rapid and decisive action. If so, the imperative of survival would then trump democracy, with its procedural dawdling, endless debate, delays, compromises, evasions, and half measures. The upshot is that under conditions of multiple and mounting stresses, some degree of authoritarianism will become increasingly attractive for many, and unavoidable should democratic governments fail to respond effectively and quickly to the demands of the long emergency.

From the work of Jared Diamond, Joseph Tainter, Thomas Homer-Dixon, and other students of societal collapse we know in great detail that societies sometimes fail catastrophically due to their inability to solve environmental problems and adapt to changes in the regional climate.[4] Jared Diamond attributes the collapse of previous societies to the failure of elites to anticipate problems, perceive existing problems, and solve known problems. And sometimes, as Diamond notes, problems can exceed the capacity of the particular society to solve them—what Homer-Dixon describes as an "ingenuity gap" (Homer-Dixon, 2000). But the collapse of earlier societies had relatively little, if any, effect on others, and so the various human experiments continued elsewhere without interruption.

Global climate change, however, is unlike the challenges that toppled previous societies. It is, after all, global, and will affect every part of the world, albeit in different ways and varying levels of severity as changes grow more pronounced. Survivors of collapsed societies could migrate elsewhere—an option less available to those in the long emergency. Moreover, because they were relatively isolated, the collapse of any one did not necessarily affect others. But in a highly connected, interdependent, and tightly coupled global society, small disturbances in one place can ripple throughout the world. Even in good times, errant financial decisions in Hong Kong can bring down long-established banks in London. Further, the levels of complexity characteristic of global society are orders of magnitude larger than those of earlier societies, as complex as they undoubtedly were. The conditions leading to the collapse of earlier societies, whether environmental or social, were in varying degrees temporary. But climate change and other aspects of the long emergency are virtually permanent as we measure time. Sea levels, for example, are expected to continue rising for the next thousand years and beyond. Finally, climate change is only one element of what futurist John Platt once described as a "crisis of crises." In short, in terms of duration, complexity, and scale, no historical precedent exists for what lies ahead.

ALTERNATIVES TO DEMOCRACY: DOWNSIZE, DESIGN, AND MARKETS (AGAIN)

There is no escaping the fact that we are entering the opening years of difficult times with no adequate political framework or philosophy. As Amory Lovins, quoted above, puts it, "We lack a theory of governance... We need to invent whole new institutions, new ways of doing business, and new ways of governing" (Gould and Hosey, 2007, p. 32). What is to be done?

One possible answer is that given by libertarians and think-tank conservatives who say that we do not need much governance in the first place and should therefore happily dispense with most of it. They propose to replace parasitical governments with "free" markets and privatize public services. Like many ideologies, this one has always worked better in theory than on Main Street. The many mistakes that led to the economic collapse of 2008, in Joseph Stiglitz's words, "boil down to just one: a belief that markets are self-adjusting and that the role of government should be minimal" (2009). Downsizing government removes any countervailing power to that of corporations and leaves the public defenseless to face, as best it can, the many problems generated by unfettered capitalism, such as unfair distribution of wealth, unequal legal protection, exposure to pollution, inadequate and expensive health care, lack of emergency help, and unavailability of basic services. Libertarian ideas appeal most strongly to those who are wealthy enough to buy their own services and live in well-defended enclaves isolated from the larger society.

A second and more realistic possibility is to reduce the need for government by redesigning energy systems, buildings, communities, manufacturing, farming, forestry, transportation, infrastructure, and waste handling in ways that mimic natural processes and radically increase local resilience. The result would be communities, societies, and eventually a global civilization running on sunshine and wind without pollution from the combustion of fossil fuels; towns and cities designed to work with natural processes; manufacturing systems that mimic natural processes and emit no pollution; and localized food systems built around sustainable farming and powered by sunshine. The agreeable result would be to eliminate the need for a great deal of environmental regulation and government interference in markets while making society more resilient in the face of climate change, oil shortages, terrorism, and economic turmoil.

In fact, a design revolution is gathering steam: solar and wind technologies are being deployed rapidly, the U.S. Green Building Council and the American Institute of Architects have adopted the goal of building only carbon-neutral buildings and communities by the year 2030, the application of biomimicry is helping a growing number of companies to eliminate waste and pollution, markets for food grown sustainably are growing and are increasingly profitable, and the technology necessary to reduce or eliminate fossil fuels is becoming available. But ecological design does not necessarily solve larger problems, such as the provision of basic services, health care, fairness, or emergency services at the scale that will be needed. And a design revolution won't go very far without a major overhaul in public policy and the tax system, as well as imaginative public investment in research and development. Ecological design, in short, is a promising step in the right direction and could reduce or eliminate many problems that perplex politicians and baffle government bureaucrats, but it cannot eliminate the need for competent government to create the larger conditions that make it possible on a societal scale in the first place.

A third, and related, possibility is to make capitalism an environmentally constructive, not destructive, force.[5] An ecologically enlightened capitalism would place value on "natural capital" such as soils, waters, forests, biological diversity, and climate stability while retaining the dynamism and creativity of markets and entrepreneurship. It would, in other words, internalize costs that were formerly off-loaded onto society and future generations, thereby eliminating pollution and waste. If this is successful, the driving forces of capitalism, such as innovation, entrepreneurship, constant change, economic growth, and even those less likeable features of greed and miserliness, will be fine as long as prices include the true costs of using natural capital. It is said that firms operating by the rules of natural capitalism inevitably will be more profitable than those that do not. Perhaps with better design, improved technology, and ecologically smarter business we don't need much government at all. But again, let's take a closer look.

Without anyone quite saying as much, the case for natural capitalism begins with the assumption that government and politics cannot be bettered and that improvement in human behavior or corporate motivation is both highly unlikely and unnecessary.

We should aim, accordingly, to harness self-interest, not loyalties to community; the power of greed, not that of altruism; and the practical force of utility but that of no larger vision. In the transition to an ecologically designed, pollution-free world operating on sunlight, there is far less for government to regulate, less cause to intervene in markets operating by the logic of natural capitalism, and fewer, if any, wars to fight over oil or resources. So what do governments do when they no longer need to regulate commerce, bother people, or fight each other? Perhaps, as Karl Marx once fantasized and tax-cutter Grover Norquist suggested more recently, the state would simply wither away.

This is an appealing vision in many ways, and on odd-numbered days I am inclined to believe some of it. But on other days I am sobered by the recollection of the lamentable history of corporations and the persistence of greed, ignorance, hard-heartedness, the lust for power, and the unfailing human capacity to screw up even good things.[6] For those prone to hyperventilate about the many virtues and sincerity of green corporations, I suggest that they spend a few hours on top of what's left of any one of the 500-plus mountains in Appalachia leveled for a few years of "cheap" coal—or perhaps sit outside the former headquarters of Enron or Lehman Brothers and contemplate what corporations have wrought to people and the land. Or they might simply ponder the Hummer—its origins and the many reasons for its demise, along with those of General Motors and the city of Detroit. Maybe a natural capitalism will be different, but we ought at least to ask why and how it would be different. Is natural capitalism as inevitable as claimed? Will it, in fact, be more profitable, as claimed? Can capitalism be rendered patient and disciplined in the long term? Will green capitalists, as a rule and not as an occasional late-life and much celebrated event, subordinate the will to accumulate to the larger good?

With few exceptions, corporate decisions to go green or reduce carbon footprints are based on the cold logic of profit in a system that is designed to maximize short-term shareholder value and in which the rights of an abstraction—the corporation—have by legal alchemy been rendered equal to those of real people.[7] There are few, if any, reasons for abstractions to protect the long-term public good, but there are many for them to appear as if they are doing so. Given the rules of the market, there is still little or no reason not to off-load environmental costs of doing business onto less-developed countries or future generations, but there are many reasons to lobby behind closed doors against rules requiring corporate accountability and decency, which are central to the vision of a natural capitalism. Even were corporations to become fully housebroken, there still might be precious little incentive for them to do their fair share to alleviate poverty or distribute income fairly, but there would be many reasons to justify not doing so as economic necessity. And there is no particular reason why they would prefer democratic government to other forms, raising the specter of a world largely controlled by solar-powered, hyperefficient, and green corporations—but that would be a sustainable form of fascism.

The theory of natural capitalism assumes that the four kinds of capital—financial, productive, ecological, and human—can be melded into a single economic framework. But, in fact, they operate by very different rules. Financial and productive forms of capital work by the laws of greed and smartness. But human and ecological forms of capital, otherwise known as people and nature, work by the laws of affection, prudence, and foresight. Advocates for natural capitalism believe that these radically different forms of "capital" can be willingly and voluntarily joined, across many different sectors of the global economy—from mining and manufacturing to services and information—in time to head off the worst of the long emergency. It is a gamble that the considerable global powers of accumulation can be voluntarily joined with the public interest in long-term sustainability without the robust convening or supervising agency of governments committed to protect the commonwealth. And it is a gamble that we can build an enduring, fair, and decent global society around smarter consumption. It is a gamble that this union will require little democratic participation in defining the public agenda or in making the decisions that affect the future of civilization. But in the absence of robust government leadership and a revitalized civic life, there is little convincing evidence that the transition to a more natural capitalism would transform enough of the global economy in time to avert disaster. There is a great deal of evidence, however, that practitioners of natural capitalism, like all previous capitalists, will make every effort to keep the consumer economy growing, come what may.

The issue is not whether it is possible for corporations to do much better, and I happily acknowledge that many are in fact doing so. Nor do I dispute the potential of better technology and improved design to reduce our carbon emissions and ecological footprint. At best, however, such things only buy us a little time to get the big things right, and those are things only governments can do: maintain a forum for public dialogue, promote fairness, resolve conflicts, provide services that markets cannot, and meet our obligations across the boundaries of politics, ethnicity, time, and species, all of which will grow more difficult in the conditions of the long emergency. Natural capitalism is a necessary but insufficient response to the long emergency ahead. But how do we revitalize democracy and our public life?

RESTORING DEMOCRACY: MEDIA, MONEY, CIVIC RENEWAL

Here the going gets harder and the issues become more contentious, but a few things, nevertheless, are obvious. The health of a democracy depends on what at least some of the founders believed to be the inherent wisdom of the people, or what James Surowiecki (2005) has more recently called "the wisdom of crowds." But the people and crowds can make bad decisions, including ones that would paradoxically destroy democracy. Foolish people, in other words, can wreck even the best possible political system, while people with more foresight and public spirit can make a lesser system work well. There is no simple remedy for public apathy, carelessness, ignorance, or meanness, but there is a steep price to be paid if such qualities become the national character. As Thomas Jefferson put it, "If a nation expects to be ignorant and free, in a state of civilization, it expects what never was and never will be." The founders, accordingly, placed their bets on an informed public, hoping "to inform their discretion by education," as Jefferson put it. Freedom of the press, accordingly, was particularly important for the authors of the Constitution. Without accurate information or the means to acquire it, in James Madison's words, political life would degenerate into "a farce or a tragedy, or perhaps both." Whether or not we have reached the level of farce or tragedy, it is clear that the press is no longer the alert watchman it once may have been and that it no longer plays the role the founders thought necessary for a healthy democracy. With a few exceptions, such as the McClatchy papers, intrepid reporters like Seymour Hersh, and columnists like Frank Rich, the late Molly Ivins, and Paul Krugman, the mainstream media and television news covered itself in ignominy in recent years, seldom challenging government statements regarding the decisions to launch the war on Iraq, violate the law by using torture, suspend the right of habeas corpus, and illegally spy on its own citizens.[8]

Part of the reason for the poor performance of the American press can be found in the increasing centralization of the media. In the first edition of *Media Monopoly* in 1983, Ben Bagdikian lamented that we were down to 50 major media outlets. When he wrote the updated version in 2005, the number had dropped to five, one of which is Fox News. Serious news and investigative reporting have been sacrificed in the competition for market share in an increasingly centralized market. When the *Millennium*

Ecosystem Assessment Report, the largest study ever done on the health of the planet, appeared in the spring of 2005, for example, none of the major news channels reported the story, and it did not make the front page of our national newspapers like the *New York Times.* Instead, the highly politicized story of a brain-dead woman (Terri Schiavo) was the lead story on the evening news and dominated most newspaper front pages as well as radio and television talk shows. The fact that the natural systems on which we depend were dying did not matter to the people who define the "news," but short-term market share did matter, and that called for sensationalism and the further cultivation of public cupidity.

There is no single solution to what is a complex problem, but obvious reforms would go a long way toward restoring the free flow of information. The first would require returning to the idea written into the Communications Act of 1934, which granted broadcasters use of public airwaves on the condition that they would serve the public interest (Hill, 2006, p. 121). But in recent years, public programming and political coverage on the major networks have plummeted along with the quantity, quality, and integrity of news reporting. The problem is simply that "the press has grown too close to the sources of power in this nation, making it largely the communication mechanism of the government, not the people" (Bennett, Lawrence, and Livingston, 2007, p.1). A second change would be to again enforce the 1947 rule by the Federal Communications Commission that required fair and balanced coverage of the news in order to hold and retain a license to use the public airwaves. Since 1987 that rule has been largely ignored by the appointees to a highly politicized FCC (Thomas, 2006, pp. 124–134). As a consequence, according to a report from the Center for American Progress, over 90 percent of talk radio across the United States, for example, is "conservative," much of it unencumbered by fact and seldom challenged in open debate. The public airwaves, in effect, have been co-opted not to inform the citizenry but to wage a "culture war" and to demonize opponents.

Exposed to a steady diet of right-wing talk radio, many Americans are misinformed about the major issues of the day, including climate change.

A third change would be to reverse the Telecommunications Act of 1996, which allowed corporations to buy up newspapers, radio, and television stations serving the same media market. The stated purpose was to encourage competition, but instead the act led to what media scholar Robert McChesney describes as "a massive wave of consolidation throughout the communications industries" (McChesney, 1999, p. 74; Bollier, 2003, pp. 148–153). Entire regions of the United States are now blanketed by highly biased and distorted news coverage provided by a single company, with no dissenting voice on the airwaves. That goes a long way to explaining why Americans are among the most media-saturated but worst informed people in the world and why we have been so confused and apathetic about impending climate destabilization. The notion of a public service is, in McChesney's view, "in rapid retreat if not total collapse" (p. 77). The International Press Association, unsurprisingly, rates the U.S. press 27th freest in the world (Gore, 2005).

Failures of the media and press, however, result from the rising tide of money that flows through the political system, corrupting everything it touches and compromising every politician, some more than others. A long time ago, Will Rogers noted that we have the best Congress that money can buy. But he hadn't seen anything. From his time to ours the situation has grown from a problem to a national disgrace, corrupting virtually every aspect of our public life, climate policy not the least. We've tried tinkering with the system to no avail. The solution, however, is straightforward: remove money from politics entirely. At the heart of the matter is the strange 1976 decision by the Supreme Court in *Buckley v. Valeo* that expenditure of money in political campaigns is a form of free speech and therefore protected by the 1st Amendment. As a result, a very few have a very large say in defining issues and electing candidates, while many have little or no say. It is time, long past time, to separate money and politics in the same way the founders intended to separate church and state. All federal elections ought to be publicly financed. The corollary is that no elected or appointed official after leaving public office should ever be allowed to hold a paid position with any regulated industry. If public officials face financial destitution as a result of their public service, let us pay them better. But the people's business should not be peddled like beer and SUVs.

Many who believe that we need a robust, democratic, and rational politics propose to harness technology in order to create an electronic version of a town meeting. After a thorough and convincing examination of the causes of the decline in the rationality in our politics, former vice president Al Gore, for one, proposes to harness the power of the internet to join us electronically as citizens across the divisions of age, geography, and ethnicity. "The internet," he argues, "is perhaps the greatest source of hope for reestablishing an open communications environment in which the conversation of democracy can flourish" (Gore, 2007, p. 260). Susan Jacoby, to the contrary, argues that the "first essential step is negative: we must give up the delusion that technology can supply the fix for a condition that, however much it is abetted by our new machines, is essentially nontechnological" (p. 309).

Another possibility is to create mechanisms in which citizens would meet in small assemblies for several days to reason through complex issues of public policy. James Fishkin and Bruce Ackerman propose, for example, a national effort to increase civic intelligence by engaging people across the political spectrum in public dialogue. A national "deliberation day" would be held every four years to coincide with presidential elections. Citizens randomly selected and paid a daily stipend would meet to discuss and debate the issues of the day, guided by rules to ensure fairness and full participation. Fishkin's early experiments indicate that the process works, at least at a small scale under careful supervision. Other evidence indicates as well that public deliberation under the right conditions does indeed increase both participation and the quality of public dialogue.

But deliberation without fair and transparent elections is meaningless. Steven Hill (2002; 2006) has written persuasively about the need to improve the electoral system by ensuring that votes are in fact counted, expanding voter participation, providing for instant runoffs, scrapping winner-take-all elections, and providing for direct election of the president, all of which are practical, widely popular, and achievable reforms lacking only the political leadership to implement them.

The future of democracy in the United States also depends a great deal on qualities that are harder to define, what de Tocqueville called "habits of the heart," which define the kind of people we are. The founders' faith in the necessity of virtue was soon diminished by reality, but stable democracy everywhere requires civil people who tolerate differences and are willing to split the difference. Can people lose the capacity for democracy? Both Madison and Jefferson thought so, and in recent decades our own democracy frayed as our political dialogue became more contentious and narrowly partisan. "Wedge issues," such as gay rights, abortion, flag burning, and (since September 11, 2001) the most strident form of patriotism, have driven out more substantive and important issues. We are said to have become a conservative country, considerable evidence to the contrary notwithstanding, including the election results of 2008 (Hacker and Pierson, 2005). Whatever the truth of the assertion, the conservatism of Rush Limbaugh, Sean Hannity, Anne Coulter, and Karl Rove has little in common with the principled conservatism of the kind once proposed by Edmund Burke, Richard Weaver, Russell Kirk, Robert Taft, and even Barry Goldwater, a conservatism that never took a firm hold in the United States. Clinton Rossiter once said that genuine conservatism was done in by twin forces of democracy (too much, too fast) and industrialism, creating a "one-way ticket to social nonconformity, financial mediocrity, and political suicide" (pp. 201–204, 212). True conservatism, as a result, "withered and died" long ago (p. 207) and descended into anger, stereotyping, sloganeering, myth-making, and "frightening simple-mindedness" (p. 209). Written a half century ago, those words were a harbinger of what was to come.

For its part, the liberalism that we associate with Franklin Roosevelt or John F. Kennedy fell victim to tragic assassinations and the Vietnam War and subsequently withered for lack of back-bone and vision. Franklin Roosevelt, still despised in some circles for introducing Social Security, was the best friend capitalism ever had; he was more of a pragmatic and creative conservative than he was a socialist. In its heyday, between 1933 and 1968, how-ever, liberalism failed to mature into a robust, agile, and sustain-able movement. To accommodate corporate interests, liberals bent with the political winds and often compromised the ideals of an open, fair, and democratic society. The public quickly smelled fear and defeatism, and if there is anything Americans don't like it is cowards and losers. Like conservatism, a full-blown liberalism never took a firm hold in America.

As a result of failures on both right and left, our recent political life and policies are strangely unrelated to the real challenges of the long emergency looming ahead. A great deal of recent "neocon-servatism" was little more than an ideological veneer to cover a hijacking of American politics by an unlikely coalition of neocon buccaneers, theocons, right-wing extremists, old-line Republicans, tax-cutters, corporations, the defense industry, and conservative evangelicals (Linker, 2007). In order to shift the public agenda, conservative donors spent an estimated $3 billion from 1970 to 2000 to create ideologically oriented think tanks and a network of radio and television stations featuring dependably angry com-mentators fulminating daily against treasonous liberals. Whatever other religious or political values might have been involved, they deflected public attention from the largest transfer of wealth in his-tory from the middle and lower classes to the extremely wealthy.

In Robert Kuttner's words: "more than half of the income lost by the bottom eighty percent was captured by the top one-quarter of one percent" (Kuttner, 2006, p.3). Not to put too fine a point on it, this was a con job sponsored by a few who had a lot to gain from public befuddlement, culture wars, and political polarization. The results were a three-decade-long, deliberately provoked public donnybrook that distracted us from more serious issues having to do with policy changes necessary to promote energy efficiency and solar power, encourage sustainable economic development, improve environmental quality, modernize transportation, and rebuild cities in order to head off climate destabilization.

In the election of 2008, a majority of Americans decided that the country could not be run indefinitely on debt, mendacity, and incompetence. But the damage to be undone is daunting, including the highest income disparity since 1929, financial markets in ruin, severe economic recession, record deficits, soaring national debt, the quagmire of wars in the Middle East, the continuing fallout from multiple corporate scandals, many federal departments and agencies in shambles, a bloated security apparatus, and a considerable loss of respect in the international community. None of this will be undone quickly. Predictably, there is an ongoing battle over the agenda of the Obama administration—whether it should be centrist or transformative in the face of the looming emergency of climate destabilization. After the election of 2008, it is apparent, too, that the Republican Party is in disarray, its moderates all but banished and its controlling conservatives increasingly isolated from mainstream public opinion and, for decades to come, burdened by the legacy of George Bush and Dick Cheney. It would be a mistake to assume that it is spent as a political force, but without significant changes in doctrine and outlook necessary to accommodate the realities of climate change, ecological limits, multiculturalism, and world opinion, it is unlikely to play a constructive part in the creation of a new order adequate to the times ahead, and that will prove to be unfortunate.

Looking back, the last few decades should teach us that democracy is vulnerable to those, whether terrorists or ideologues of any sort, who flagrantly defy the rules of civility, tolerance, and public order. The history of Greek democracy, again, stands both as a beacon to the possibilities of self-governance and a warning about its fragility. Looking to the future, ours will one day appear as an oddly disoriented time. Many of the issues that fueled the passions of our day will appear to them as merely vaporous diversions from much larger issues. In particular, our obsession with consumption and individual rights to the neglect of collective rights will appear derelict, perhaps criminally so.

BEYOND LEFT AND RIGHT: THE CASE FOR PROTECTING POSTERITY

We are at the end of an age of isms—socialism, Marxism, and capitalism—all of which in varying ways held that economic growth and technology could solve all of our problems. The 18th-century Enlightenment belief in the possibility of human improvement has been whittled down to little more than the hope for continual material betterment, which in turn is threatened by ecological and demographic realities, and by our own psychology.[9] Only the true believers and a few neoclassical economists remain on the crumbling ramparts of paradigms lost. But the need for orienting principles, durable political ideas, and practical visions that join us across old partisan divisions is greater than ever. I do not think the prospect of following such principles is as unlikely as it may first seem, however, because we've done it before. In the late 1960s, for example, Republicans and Democrats assimilated the evidence about environmental deterioration and joined to create the National Environmental Policy Act (1969), the Clean Air Act (1970), the Clean Water Act (1972), and the Endangered Species Act (1973). Preserving and enhancing the environment was widely regarded as central to the national interest, not dismissed

---

Box 2.1.

Some conservative intellectuals are beginning to recognize the seriousness of climate destabilization. A notable example is that of Richard Posner, a judge of the U.S. Court of Appeals for the Seventh Circuit and a prolific and often brilliant writer on the law, literature, and politics. He is perhaps best known as an advocate for the use of economic standards as both a legal tool and a yardstick by which to measure judicial decisions. Judge Posner advocates wealth creation as a reliable standard for legal reasoning, along with great deference to prevailing practices and behaviors. Applied to politics, Posner's pragmatism comes close to views once proposed by the economist Joseph Schumpeter, himself no flaming reformer. Posner's position has been so forcefully and consistently stated for so long that it is surprising to read his book *Catastrophe* (2004), in which he concludes that the odds of one or more catastrophes are growing quickly. Among these he includes the prospect of rapid climate change and admits that it "is to a significant degree a by-product of the success of capitalism in enormously increasing the amount of world economic activity . . . and is a great and growing threat to anyone's idea of human welfare" (2004; p. 263). On this subject, conservatives, he believes, are "in a state of denial." The problem has come about, in part, because of the "scientific illiteracy of most nonscientists . . . [particularly] the people who count in making and implementing policy" (p. 264).

    Posner believes that the dangers of one or more catastrophes are growing because of "the breakneck pace of scientific and technological advance" (p. 92). As for a framework to understand our situation and to reduce the potential for disaster, the "natural candidate . . . is economics," but alas, the subject of catastrophe "turns out to be an unruly subject for economic analysis" (p. 123). This is so, in some measure, because of the global scope of the problems and because of the long-time horizons involved, which bring into question standard economic tools such as discounting. The use of discounting thus carries the disadvantage of raising the potential for disasters that occur sufficiently far into the future

(continued)

Box 2.1. (*continued*)

that the benefits of procrastination and the costs of disaster fall onto different generations without the possibility of some offsetting benefit by the accrual of more wealth. If economics is an unsatisfactory tool, the law is little better. Indeed, Posner believes that "the legal profession may even be increasing the probability of catastrophe" (p. 199). Improvement in this situation, in his view, will require "that a nontrivial number of lawyers" become scientifically literate, an interesting challenge (p. 203). Posner further proposes other remedies, such as the establishment of a science court, a center for catastrophic-risk assessment, the use of fiscal tools such as taxation and subsidies, increased regulation including the establishment of an international EPA, increased scrutiny of research projects in high-risk areas, and greater police powers to detect and control growing risks of terrorism.

Posner's recent concern about the rising potential for catastrophe is welcome and significant. But it calls for some explanation. His voice and considerable influence have long been on the other side of environment-related issues, encouraging his many influential readers to regard the market and wealth creation as the primary standard for the law and public policy. He is a member of the University of Chicago group, including Milton Friedman and Richard Epstein, who gave us the prevailing economic and legal philosophy that prized individual rights and the free-market ideology that have destroyed a great deal of our capacity for public responses to public problems. Nonetheless, it will be important for legal scholars to sort out issues of law and liability in climate-change cases that resemble the tobacco cases. The human and property consequences of the use of fossil fuels were known well in advance of efforts to reduce or mitigate them. As with the tobacco cases, scientists have been warning us at least since the late 1970s in ever more insistent terms, and no one can ever legitimately plead that they did not know the consequences of their actions.

as merely partisan opinion. Americans, too, joined across party lines to overcome the Great Depression of the 1930s and combat fascism and communism. Americans have risen to meet daunting challenges in the past, and we can do it again.

A shared national agenda that joins the public across the political spectrum, however, cannot be built on outworn myths and historical illusions. Our survival and that of our democracy depend on a clearer understanding of our own ecological history and the recognition of our disproportionate role in causing rapid climate change. We have often thought ourselves to be a unique and blessed people, and in some ways we may be, but a great deal of that uniqueness is owing to the fact that our forefathers discovered the blessings hidden in the last and greatest reserve of stored carbon on Earth. Our soils and forests were some of the richest anywhere, and our supplies of coal and oil seemed inexhaustible. And rather like yeast cells feeding on sugar in a wine vat, we prospered by exploiting carbon and depleting soils, forests, and fossil fuels alike. As a result, we Americans release an average of 22 tons of CO_2 per person each year and are responsible for 28 percent of the increased carbon in the atmosphere, but we make up only 5 percent of world population. In time our good fortune gave rise to an inordinate sense of self-congratulation and the belief that, so endowed, we must be God's favored people. From there it was but a short step to intoxicating doctrines of manifest destiny and later to a foreign policy built on the idea of American supremacy. But the access to cheap and abundant carbon led also to excesses of overconsumption and waste, an epidemic of fatness, urban sprawl, violence, and energy profligacy, and lifestyles that can be neither sustained here nor duplicated for very long elsewhere. Historian David Potter once characterized Americans as "a people of plenty," but in time we became a people of excess, and we now have to find our way back to older values of thrift, frugality, neighborliness, and what John Todd once called "elegant solutions predicated on the uniqueness of place." That is a homecoming of sorts.

Our future and the future of our democracy depend greatly on our sense of connection to each other and our obligations to future generations. Economist Kenneth Boulding once asked in jest "what has the future done for me…lately?" By definition the unborn can do nothing for the living, but the idea of posterity—a decent future for our children and theirs—does a great deal for us. Our hopes for the future inspire the best of which we are capable. Absent that hope, we easily fall victim to shortsightedness, anomie, and carelessness in managing our contemporary affairs. Even were that not so, are we in some manner obliged to care about generations to come? The founder of modern conservatism, Edmund Burke, once described the living generation as trustees obligated to pass the inheritance of civilization from the distant past on to future generations. That inheritance includes all the best of civilization—our laws, customs, culture, and institutions—and the ecological requisites—clean air and water, healthy ecosystems, and stable climate—on which they depend. But without advocates in the present and lacking any standing in the law, posterity has no power to enforce its rights. The U.S. Constitution mentions posterity in the preamble but not thereafter. In the more than two centuries since the Constitution was written, no significant case law has developed to protect posterity, leaving it defenseless against harms perpetrated on it, knowingly or not, by previous generations.[10] It might be argued that it has always been the case that each generation benefits from the progress bequeathed by earlier generations and suffers the effects of, say, soil loss or the loss of biological diversity accidentally incurred. It is presumed, however, that on balance the benefits more than offset the losses. It is an open question how much the members of any generation know the effects of their actions on later generations. But until roughly the mid-20th century, the scale of costs imposed from one generation to the next was contained locally or regionally, and the damage was often repairable in a matter of decades or centuries. The intergenerational costs of climate change, however, are another matter entirely. They are global, permanent (as we measure time), and mostly well understood. We cannot plead ignorance about the facts of climate change or argue that time will heal the damages we cause. And, given the large and well-documented evidence of the potential for energy efficiency and renewable energy, neither can we make a plausible case that we had no other choices. We will stand before whoever is able and willing to judge, or perhaps the silence of extinction, as a generation that willfully and unnecessarily imposed egregious wrongs on all future generations, depriving them of liberty, property, and life. We are culpable, but the law as presently constituted conveniently lets us off the hook because it says nothing about the rights of posterity.

Harms perpetrated across generations or actions that could lead to the extinction of humankind are perhaps the most baffling, as well as most important, issues of ethics and policy. In the latter case, for instance, the unborn are harmed only in the sense that they were not brought into existence, which raises many perplexities. Pondering self-inflicted extinction, Jonathan Schell, for example, writes:

> How are we to comprehend the life or death of the infinite number of possible people who do not yet exist at all? How are we, who are a part of human life, to step back from life and see it whole, in order to assess the meaning of its disappearance? To kill a human being is murder.... but what crime is it to cancel the numberless multitude of unconceived people? In what court is such a crime to be judged? Against whom is it committed? And what law does it violate? (Schell, 2000, p. 116)

Regarding the rights of future generations, Schell asks, "What standing should they have among us?" They exist nowhere except in prospect, and the law presently affords no protection or consideration to humans "unconceived." Schell was writing about the case of extinction by nuclear weapons, but extinction by climate change poses virtually the same difficult issues, albeit less quickly.

At a slightly less vexatious level, the case against abortion is perhaps instructive. Whatever one's opinion about the rights of a fetus, might the same kind of arguments apply to the right to life of future generations? Some will say that rights can be assumed only when agents are able to reciprocate, hence they must be living at the same time and capable of reciprocity. But we provide for the unborn in many ways without making any such assumptions. Some people endow colleges, universities, and cultural institutions for future generations. From Thomas Jefferson to the present it has been common to object to the imposition of our financial burdens on posterity (Yarrow, 2008). In this case, however, it is the right of future generations to life itself.

James Madison gave us another way to understand intergenerational obligation. In the tenth of the *Federalist Papers,* he made the case for controlling the power of factions in order to protect the larger good. But from the perspective of posterity, the present generation is but a faction with the unchecked power to consume resources and inflict irreparable damage to all subsequent generations. Madison could not have foreseen the kinds of harm one generation could inflict on subsequent generations, but the logic is equally clear for controlling the power of factions in both instances.

Legally and morally, however, we are in new territory, and we struggle to find the concepts and words to describe our situation in the hope that being able to name it, we might avert it. During World War II the word "genocide" was coined to describe the systematic and willful destruction of entire ethnic groups. But we have no word to describe our own actions, the consequences of which are now killing what the World Health Organization estimates to be 150,000 people each year and will cause the death of millions more in the future, and perhaps much worse. The effects of our present use of coal, oil, and natural gas will kill into the far future, but we cannot know exactly who, where, or how they will die. We do know, however, that the number will be very large and that they will perish in storms, or heat waves, or of strange diseases, or in violence amplified by famine, or in any of a thousand other ways. We have, however, no word by which to describe calamity at this scale and, as yet, no means to hold perpetrators accountable. It is rather like Churchill's description of genocide as "a crime without a name," but on an infinitely larger scale.[11]

Other than those in the present willing to speak and act on their behalf, generations to come have no defenders. In contrast to the fetus, they exist only in prospect, but like a fetus, that prospect can be radically crippled or aborted by the indifference or dereliction of the present generation. And in contrast to the individual fetus, future generations pose a collective challenge both to the law and to our capacity to make moral decisions. To do so, we must imagine their lives at a scale for which we are unaccustomed and must summon the wherewithal to act on their behalf as well as the intellectual acuity to know how to act effectively to defend their interests. Since it is within our power to grant or to withhold life, what can we do?

Any posterity capable of making such judgments will regard the policy and behavior changes necessary now to limit and reverse the damage of climate change to them as obvious and necessary. The real fault line in American politics is not between liberals and conservatives or between Democrats and Republicans. It is, rather, in how we orient ourselves to the generations to come, who will bear the consequences, for better and for worse, of our actions. Some say that we cannot know precisely what the consequences of rapid climate change may be and that we should not act until they are known in great detail. Others, mostly devotees of the faith of neoclassical economics, say that we cannot afford to act and must wait until we are richer still. There are still a few who believe that climate change is a hoax and prefer to do nothing— come what may. There are many, perhaps still a majority, who will say nothing because they have not thought much about such things. But a growing number believe that we must act now.

The long emergency will be a test of our beliefs, institutions, and character as a people. We all hope fervently that we and our children and theirs will come through what biologist E. O. Wilson calls "the bottleneck" with a more tempered but improved and more vital democracy. But I do not believe that to be even remotely possible without extending our boundaries of consideration and affection to include posterity and, in some manner yet unknown, other life forms. I think this is what Albert Einstein was getting at when he proposed that:

> A human being is part of the whole world, called by us "Universe," a part limited in time and space. He experiences himself, his thoughts and feelings as something separate from the rest—a kind of optical delusion of his consciousness.

Aldo Leopold, similarly, wrote in *A Sand County Almanac* that "a land ethic changes the role of *Homo sapiens* from conqueror of the land-community to plain member and citizen of it." Both Einstein and Leopold regarded our diminished sense of community as part of the unfinished business of human advancement. And both regarded this development not as a burden so much as growth in our human stature. It is time to expand our political horizons in ways commensurate with the extent of our effects on the future and the community of life. It is time to grant standing to our posterity, whose lives, liberty, and property are imperiled by our actions, and include them in a larger democracy. The principle involved draws from our own revolutionary experience and could be simply stated as:

> No generation and no nation has the right to alter the biogeochemical cycles of Earth or impair the stability, integrity, or beauty of natural systems, the consequences of which would fall as a form of intergenerational remote tyranny on all future generations.

This wording draws from Thomas Jefferson and the generation that threw off the arbitrary authority of a king, Aldo Leopold's description of a morally and ecologically solvent land ethic, and hundreds of contemporary ethicists and scientists who have wrestled with the darkening shadow that our generation casts onto succeeding generations and the opportunities we have to lighten that darkness.

Box 2.2. Postscript: A Note on the Shelf Life of Economic Ideas

Those of us who have looked to the self-interest of lending institutions to protect shareholder's equity, myself included, are in a state of shocked disbelief... Yes, I have found a flaw. I don't know how significant or permanent it is. But I've been very distressed by that fact.

—Alan Greenspan

The self-confidence of learned people is the comic tragedy of civilization.

—Alfred North Whitehead

Conservative philosopher Richard Weaver once noted that "ideas have consequences" (1984). And some really bad ideas of the last half century are leaving a legacy of very bad consequences. Weaver's 1948 book was an extended argument for conservatism, beginning with the recognition of knowledge higher than our own and the importance of such things as virtue, character, craftsmanship, enduring quality, civility, and, above all, piety. Applied to nature, Weaver argued for a "degree of humility" such that we might avoid meddling "with small parts of a machine of whose total design and purpose we are ignorant" (p. 173). "Our planet," he wrote, is falling victim to a rigorism, so that what is done in any remote corner affects—nay, menaces—the whole. Resiliency and tolerance are lost" (p. 173). Weaver regarded the modern project to reconstruct nature as an "adolescent infatuation." One can reasonably imagine the approbation he would have felt for the creative exhibition of thievery and stupidity that has led to our present circumstances.

Weaver's idea that ideas have real consequences, alas, had less consequence than one might wish. It is honored mostly among a small band of true conservatives, the uncommon sort who actually value the conservation of tradition, law, custom, nature, culture, and religion, and who take ideas and their real-world implications seriously. Other than the title of his book, however, Weaver is presently unknown to the wider public, and probably not at all to the faux conservatives who daily bloviate on FOX News. Unfortunately, ideas, whatever their consequences, seldom "yield

(continued)

BOX 2.2. (continued)

to the attack of other ideas," in John Kenneth Galbraith's words, "but to the massive onslaught of circumstances with which they cannot contend" (2001, p. 30). That appears to be true in our own time, in which the pecuniary imagination was given such full reign. The convenient idea that foxes could be persuaded to reliably guard the henhouse, for example, derived from free marketeers like Milton Friedman, libertarians like George Gilder, supply-side economists like Arthur Laffer, and "long-boomers" like Peter Schwartz, did not voluntarily surrender to superior reason, logic, or evidence. Rather, it was an idea whose consequences turned out to be bad for both the hens and a bit later for the starving foxes, some of whom, now professing different ideas, stand in line for public bailouts—roadkill on the highway called reality. The shelf life of such ideas will turn out to be brief as such fads go, but the consequences will last a long time.

When delusion is popular, however, durable ideas are unpopular, or more likely forgotten altogether. But in the present wreckage we have no choice but to search for more durable ideas with more benign or even positive consequences. When we find truly durable ideas, they are mostly about limits to what we can do or should do—but restraint, prudence, and caution are, "oh mah God, sooo not cool" as one of my students thoughtfully expressed it. Accordingly, such things are put on the shelf, where they gather dust until necessity strikes again and they are called back into use as we try once again to find our bearings amidst the debris of popular delusions gone bust.

In this regard, an ancient collection of proverbs contains these words of the Greek poet Archilochus: "the fox knows many things; the hedgehog knows one big thing," Like the hedgehog, advocates for the environment, animals, biological diversity, water, soils, landscapes, and climate stability know one big thing, as biologist Garrett Hardin once put it, which is that "we can never do merely one thing" (Hardin, 1972, p. 38). In other words, there are many unforeseen consequences from what we do, and so there are limits to what we can safely do. Since consequences are not only unpredictable but often remote in time and distant from the cause, we are often ignorant of the victims of our actions, and so there are moral limits on what we should do as well. To think about consequences over time requires, further, that we know how things are linked as systems and understand that small actions can have large consequences, many of which are unpredictable.

Around the first Earth Day in 1970, there was an efflorescence of brilliant thinking along these lines. In different ways, it was mostly about the things we could not do. Rachel Carson's *Silent Spring* (1962), for example, launched the modern environmental movement with the simple message that we could not carelessly spread toxic chemicals without causing damage to animals and eventually to ourselves. The book was attacked by proponents of what she called "Neanderthal biology," most of whom—then and now—with a great deal of money and/or reputation invested in the petrochemical business.

Other books making the same point addressed unlimited growth of population, economies, technology, and scale. However prescient and true, most of the wisdom was quickly forgotten. Now, however, we live in "the age of consequences" and have good reason to rethink many ideas, and the systems of ideas called paradigms. Perhaps this is what educators describe as a "teachable moment" and inventors refer to as the "aha" moment. Assuming that it may be so, I have some suggestions for those assigned to rebuild the U.S. and global economies.

The first is the old idea that we cannot build a durable economy that is so utterly dependent on trivial consumption. In 2007, for instance, Americans spent $93 billion on tobacco and another $83 billion on casino gambling, but only $46 billion on books. Another example is from the recent SkyMall catalog found in the seat pocket of commercial airplanes, which announces that it is "going beyond the ordinary." To do so, it offers those burdened with money and credit such items as a "startlingly unique" two-foot-high representation of Big Foot, to be placed in the garden where it will no doubt amaze and delight, available for only $98.95. "The keep your distance bug vacuum," equipped with a 22,400-rpm motor, is available for $49.95. And for just $299.99 cat

(continued)

Box 2.2. (continued)

lovers can buy a marvel of advanced technology: "The 24/7 self-cleaning, scoopfree litter box!" Technology-oriented catalogues regularly offer dozens, nay, hundreds of devices that digitally amaze, ease, simplify, gratify, sort, store, scratch, waken, warn, multiply, compute, freshen, check, sanitize, and personalize. It may be possible, one day, to live in a digital, stainless steel nirvana of the sort George Orwell once said would "make the world safe for little fat men."

There are, however, many problems with an economy so dependent on ephemeralities. It is a cheat because it cannot satisfy the desires that it arouses. It is a lie because it purports to solve by trivial consumption what can only be solved by better human relations. It is immoral because it takes scarce resources from those who still lack the basics and gives them to those with everything who are merely bored. It is unsustainable because it creates waste that destroys climatic stability and ecosystems. It is unintelligent because it redirects the mental energies of producers and consumers alike to illusion, not reality, which makes us stupid. And because of such things an economy organized to promote fantasy will eventually collapse of its own weight.

In *The Memory of Old Jack*, Wendell Berry describes the main character as "troubled and angered in his mind to think that people would aspire to do as little as possible, no better than they are made to do it, for more pay than they are worth." The masters of the recently imploded financial universe who made millions while destroying much of the economy, including the chief executive officers of any number of corporations from Enron to General Motors, would have appropriately aroused Old Jack's fury, as it should ours.

I have a second suggestion, which is simply that we ought to build a slower economy. It is also an old idea, embodied in aphorisms such as "the race is not to the swift" and "haste makes waste." In the age of hustle, cell phones, and instant everything, we are inclined to forget that lots of worthwhile things can only be done slowly. It takes time for all of us, economists included, to think clearly. It takes time to be a good parent or friend. It takes time to create quality. It takes time to make a great city. It takes time to restore soil. It takes time to restore one's soul. In each case, speed distorts reality and destroys the harmonies of nature and society alike (Orr, 1998). Hurry certainly changes society for the worse.

Ivan Illich's provocative 1974 book *Energy and Equity* makes the case that "high quanta of energy degrade social relations just as inevitably as they destroy the physical milieu" (1974, p. 3). "Beyond a critical speed," Illich argued, "no one can save time without forcing another to lose it" (p. 30). But any proposal to limit speed "engenders stubborn opposition... expos[ing] the addiction of industrialized men to consuming ever higher doses of energy" (p. 55). Even assuming nonpolluting energy sources, the use of massive energy of any sort "acts on society like a drug that is... psychically enslaving" (p. 6). The irony, Illich says, is that the time it takes to earn the money to travel at high speed divided into the average miles traveled per year gives a figure of about 15 mph... about the average speed of travel in the year 1900, but at considerably higher cost.

What does a slow economy look like? Woody Tasch, chairman of Investors' Circle, offers one view: "It would be driven by... the imperatives of nature rather than by the imperatives of finance. Its first principle would be, I suppose, the principle of carrying capacity, embedded in a process of *nurturing*" (2008, p. 175). A slow money economy would change the way we invest and discipline the expectations of quick returns to capital to, say, 5 to 8 percent per year, which now sounds pretty good. It would require buyers, for example, to hold stock for, say, six months before they could sell. Tasch calls this "patient capital," but by any name it involves the recalibration of money and finance to the pace of nature. And that would be a revolution.

My third suggestion is to build an economy on ecological realities, not on the belief that we are exempt from the laws of ecology and physics. In his classic 1980 book *Overshoot*, William Catton writes "The alternative to chaos is to abandon the illusion that all things are possible" (p. 9). He

(continued)

Box 2.2. (continued)

goes on to say, "We need an ecological worldview; noble intentions and a modicum of ecological information will not suffice"
(p. 12). Each ratchet upward of human population and dominance
required the diversion of "some fraction of the earth's life-support-
ing capacity from supporting other kinds of life to supporting our
kind" (p. 27). Eventually, the method of enlarging our estate by
expanding into unoccupied lands gave way to industrialization and
drawing down ancient ecological capital. "The myth of limitless-
ness dominated people's minds" to the point where we have nearly
trapped ourselves (p. 29). For Catton, we are caught in an irony of
epic proportions: "The very aspect of human nature that enabled
Homo sapiens to become the dominant species in all of nature was
also what made human dominance precarious at best, and perhaps
inexorably self-defeating" (p. 153).

Nobel Prize–winning chemist and economic theorist Frederick Soddy made a similar point in *Wealth, Virtual Wealth, and Debt*
(1926), arguing that: "Debts are subject to the laws of mathematics
rather than physics. Unlike wealth, which is subject to the laws
of thermodynamics, debts do not rot with old age and are not
consumed in the process of living. On the contrary, they grow at
so much per cent per annum, by the well-known mathematical
laws of simple and compound interest" (quoted in Daly, *Beyond
Growth,* 1996, p. 178).

"Debt," in Herman Daly's words, "can endure forever;
wealth cannot, because its physical dimension is subject to the
destructive force of entropy" (p. 179). As a result, Daly continues, "The positive feedback of compound interest must be
offset by counteracting forces of debt repudiation, such as inflation, bankruptcy, or confiscatory taxation, all of which breed
violence" (p. 179). The growth of the money economy in reality represented the expansion of claims (debt) against a stable
or now diminishing stock called nature. As the economy grew,
what we call wealth represented only a growing number of
claims against a finite stock of soil, forests, wildlife, resources, and
land, and hence was the source of long-term inflation and ruin.

Although he apparently did not know of Soddy's work, economist Nicholas Georgescu-Roegen later made many of the same points about the relation of entropy to economic growth in his monumental but widely ignored *The Entropy Law and the Economic Process* (1971). "Every Cadillac produced at any time," he wrote, "means fewer lives in the future." Given our expansive nature and the laws of physics, our fate, he concluded, "is to choose a truly great but brief, not a long and dull, career" (p. 304). Or as John Ruskin once put it more poetically: "the rule and root of all economy—that what one person has, another cannot have; and that every atom of substance, of whatever kind, used or consumed, is so much human life spent" (p. 192).

We are running two deficits simultaneously, and we must solve them together. If we fail to do so, nature will take its course. This will require a great deal of rethinking, and it will not be easy. But the present economic collapse is too far-reaching and the threat of climate disaster too real to do otherwise. Taken together, they indicate that we are nowhere near as rich as we once presumed. We have been living far beyond our means by drawing down natural capital, rather like a corporation selling off assets in a fire sale and calling the proceeds profit. The housing bubble, dishonest accounting, and the use of unaccountable financial instruments like derivatives are merely the tip of a far larger problem that includes the failure to account for carbon emissions and the loss of species diversity.

The ideas that lead us to the brink are the equivalent of junk bonds and derivatives, unsecured by real assets and ungrounded in reality. There are better ideas by which to order our economic and ecological affairs, based on the principle that "for every piece of wise work done, so much life is granted; for every piece of foolish work, nothing; for every piece of wicked work, so much death" (Ruskin, p. 202).

CHAPTER 3

Leadership in the Long Emergency

Here's the really inconvenient truth: We have not even begun to be serious about the costs, the effort and the scale of change that will be required to shift our country, and eventually the world, to a largely emissions-free energy infrastructure over the next 50 years.

—Thomas Friedman ("The Power of Green," 2007)

IN JUNE OF 1858 ABRAHAM LINCOLN BEGAN HIS ADDRESS at Springfield, Illinois by saying, "If we could first know *where* we are, and *whither* we are tending, we could then better judge *what* to do, and *how* to do it." He spoke on the issue of slavery that day with a degree of honesty that other politicians were loath to practice. At Springfield he asserted that "A house divided against itself cannot stand...this government cannot endure, permanently half *slave* and half *free*." His immediate targets were the evasions and complications of the Kansas-Nebraska Act of 1854 and the Supreme Court ruling handed down in the Dred Scott case, but particularly those whom he accused of conspiring to spread slavery to states where it did not already exist. In his speech Lincoln accused Senator Stephen Douglas, President Franklin Pierce, Supreme Court Justice Roger Taney, and President James Buchanan of a conspiracy to spread slavery. This accusation was supported by circumstantial evidence such that it was "impossible to not *believe* that Stephen and Franklin and Roger and James all understood one another from the beginning, and all worked upon a common *plan* or *draft* drawn up before the first lick was struck." His opponent in the upcoming Senatorial election, Stephen Douglas, he described as a "caged and toothless" lion.

Lincoln had begun the process of "framing" the issue of slavery without equivocation, but in a way that would still build electoral support based on logic, evidence, and eloquence. On February 27, 1860, Lincoln's address at the Cooper Institute in New York extended and deepened the argument. He began with words from Stephen Douglas: "Our fathers, when they framed the Government under which we live, understood this question just as well, and even better, than we do now." He proceeded to analyze the historical record to infer what the "fathers" actually believed. Lincoln in a masterful and lawyerly way identified 39 of the founders who had "acted on the question" of slavery in decisions voted on in 1784, 1787, 1789, 1798, 1803, and 1820. In contrast to the position held by Douglas, Lincoln showed that 21 of the 39 had acted in ways that clearly indicated their belief that the federal government had the power to rule on the issue of slavery and that the other 16, who had not been called upon to act on the issue, had in various ways taken positions that suggested that they would have concurred with the majority.

Having destroyed Douglas's position that the federal government lacked authority to act, Lincoln proceeded to address "the Southern people... if they would listen." He began with the assertion that every man has a right to his opinion, but "no right to mislead others, who have less access to history and less leisure to study it," and proceeded down the list of charges and counter-charges in the overheated politics of 1860. His aim was to join the Republican cause with the constitutional power to restrain the extension of slavery and not to assert the power of the federal government to abolish it, while also saying bluntly that slavery was wrong. He admonished his followers to "calmly consider [the] demands" of the Southern people and "yield to them if, in our deliberate view of our duty, we possibly can." And then he closed by saying "*Let us have faith that right makes might, and in that faith, let us, to the end, dare to do our duty as we understand it*" [emphasis in Lincoln's text]. Lincoln had defined the issue and clarified the powers of the federal government to deal with it.

The Cooper Institute address was instrumental in Lincoln's election to the presidency and also in framing the constitutional issues over slavery and states' rights that had smoldered for 74 years before bursting into the firestorm of the Civil War. As president, Lincoln further refined the issues of slavery, states' rights, and constitutional law. In his first inaugural address, in 1861, Lincoln attempted to reach out to the "people of the Southern states," assuring them that he neither claimed nor would assert a right as president to "interfere with the institution of slavery in the States where it exists." The address is an extended description of the constitutional realities as Lincoln saw them, in which a national government could not be dissolved by the actions of the constituent states. The point was that the Union remained unbroken and that he'd sworn only to defend the Constitution and the union that it had created, not to abolish slavery. He regarded himself still as the president of the southern states and the conflict as a rebellion, not a war between independent countries. Lincoln admonished his "countrymen" to "think calmly and *well*" on the issues at hand and then closed with the words:

> I am loth [*sic*] to close. We are not enemies, but friends. We must not be enemies. Though passion may have strained, it must not break our bonds of affection. The mystic chords of memory, stretching from every battle-field, and patriot grave, to every living heart and hearthstone, all over this broad land, will yet swell the chorus of the Union, when again touched, as surely they will be, by the better angels of our nature.

The call went unheeded and war came.

Through the next four years, Lincoln continued to frame the meaning of the Civil War relative to the Constitution, but always in measured strokes, looking to a horizon that most could not see. The Emancipation Proclamation was carefully fitted to the war situation of 1862, and the nuances of keeping the loyal slave states neutral proclaimed only a partial emancipation applicable only to the states in rebellion, drawing the ire of the impatient. At Gettysburg, Lincoln, in a masterpiece of concise eloquence based on years of arguing the principle that "if men are created equal, they cannot be property," corrected the Constitution, in Garry Wills' view, without overthrowing it (Wills, 1992, pp. 120, 147). The "unfinished work" he described was that of restoring the Union and, in effect, taking a country of states to "a new birth of freedom" as a nation with a government "of the people, by the people, for the people."

Lincoln's second inaugural address is the capstone of his efforts to frame slavery and the Constitution, and describe a nation dedicated to the proposition that all men are created equal. The setting was the final months of the Civil War, with Confederate armies on the threshold of defeat. Lincoln's tone is somber, not triumphal. While both sides in the war prayed to the same God, the prayers of neither were answered in full. "The Almighty," Lincoln reminds the nation, "has His own purposes," which transcend those of either side in the war. Drawing from Matthew 7:1, Lincoln cautioned the victorious not to judge former slaveholders, "that we be not judged." He closes by saying, "With malice toward none; with charity for all…let us…bind up the nation's wounds; to care for him who shall have borne the battle, and for his widow, and his orphan—to do all which may achieve and cherish a just and lasting peace."

From his earliest statements on slavery to "charity for all," Lincoln progressively framed the issues of slavery in ways that left no doubt that he thought it was a great wrong but preservation of the Constitution was the prior consideration. When war came,

Lincoln's first aim was to maintain the Union, but he then used the occasion to enlarge the concept of a "nation conceived in liberty and dedicated to the proposition that all men are created equal."

We are now engaged in a global conversation about the issues of human longevity on Earth, but no national leader has yet done what Lincoln did for slavery and placed the issue of sustainability in its larger moral context. It is still commonly regarded, here and elsewhere, as one of many issues on a long and growing list, not as the linchpin that connects all of the other issues. Relative to the large issues of climate change and sustainability, we are virtually where the United States was, say, in the year 1850 on the matter of slavery. On the art of framing political and moral issues, much (perhaps too much) of late has been written (Lakoff, 2004). But Lincoln did not frame the issue so much as he established a moral position on slavery that he explained in great detail. What can be learned from Lincoln's example?

First, Lincoln did not equivocate or agonize about the essential nature of slavery. He did not overthink the subject; he regarded slavery as a great wrong and said so plainly and often. "If slavery is not wrong," he wrote, "nothing is wrong." Moreover, he saw the centrality of the issue to other issues on the national agenda, such as the tariff, sectionalism, and national growth. Second, more clearly than any other political figure of his time, he understood the priority of keeping the constitutional foundation of the nation intact and addressing slavery within the existing framework of law and philosophy. He did not set out to create something from whole cloth, but built a case from trusted sources ready at hand. Third, he used language and logic with a mastery superior to that of any president before or since. Lincoln was a relentless logician, but always spoke with vernacular eloquence in words that could be plainly understood by everyone. Fourth, while the issue of slavery was a great moral wrong, Lincoln did not abuse religion to describe it. While his language was full of

Biblical metaphors and allusions, he avoided the temptation to demonize the South and to make the war a religious crusade. Throughout the seven years from the House Divided speech to his assassination in 1865, Lincoln's presidency is one of the masterpieces of transformative leadership, combining shrewdness and sagacity with moral clarity. The result was a progressive evolution of a larger concept of the nation.

From Lincoln's example we might learn, first, to avoid unnecessary complication and contentiousness. Climate change and sustainability are primarily issues of fairness and intergenerational rights, not primarily ones of technology or economics, as important as these may be. Lincoln regarded slavery as wrong because no human had the right to hold property in another human being, not because it was economically inefficient. He did not say that the country should abolish slavery because doing so would help business make more money. Rather, he said it should be abolished because it was wrong, and that moral clarity was the magnetic north by which he oriented his politics. By a similar logic, ours is in the principle that no human has the right to diminish the life and well-being of another, and no generation has the right to inflict harm on generations to come. Lincoln did not equivocate on the issue of slavery, nor should we on the tyranny one generation can now impose on another by leaving it ecologically impoverished. Climate change and biotic impoverishment are prime examples of intergenerational remote tyranny and as such constitute a great and permanent wrong, and we should say so. Each generation ought to serve as a trustee for posterity, a bridge of obligation stretching from the distant past to the far future. In that role each generation is required to act cautiously, carefully, and wisely (Brown, 1994). In Wendell Berry's words, this "is a burden that falls with greatest weight on us humans of the industrial age who have been and are, by any measure, the humans most guilty of desecrating the world and of destroying creation" (2005, p. 67).

Lincoln built his case from sources familiar to his audience—the Declaration of Independence, the Constitution, and the Bible. In doing so, he took Jefferson's views on equality to their logical conclusion and recast the Constitution as the foundation for a truly more perfect union that could protect the dignity of all human beings. In our time we can draw on similar sources, but now much enhanced by other constitutions and laws and proclamations of the world community. The Universal Declaration of Human Rights and the Earth Charter, for example, describe an inclusive political universe that extends a moral covenant to all the people of Earth and all those yet to be born. It is reasonable to expand this covenant to include the wider community of life, as Aldo Leopold once proposed.

Lincoln's use of religion is instructive both for its depth and for its restraint. He used Biblical imagery and language frequently, but did not do so to castigate Southerners or to inflate Northern pretensions. His use of religion was cautionary, aimed to heal, not divide. Lincoln oriented the struggle over slavery in a larger vision of an imperfect nation striving to fulfill God's justice on earth. The message for us is to ground the issues of climate change and sustainability in higher purposes resonant with what is best in the world's great religions but is owned by no one creed.

Because he understood the power of language to clarify, motivate, and ennoble, Lincoln used words and verbal imagery more powerfully and to better effect than any other president. This wasn't what we now call "spin" or manipulation of the gullible, but the art of persuasion at its best. Lincoln had neither pollsters to tell him what to say nor speechwriters to create his message and calibrate it to the latest polls. He wrote his own addresses and letters, and is reported to have agonized sometimes for hours and days to find the right words to say clearly what he intended. He spoke directly, often bluntly, but softened his speeches with humor and the adroit use of metaphor and homespun stories. The result was to place the horrors of combat and the bitterness of sectional strife into a larger context that motivated many to make heroic sacrifices and a legacy of thought and words that "remade America," as Garry Wills puts it. Now perhaps more than ever we turn to Lincoln for perspective and inspiration.

The tragedy of the Civil War originated in the evasions of the generations prior to 1861. The founders chose not to abolish slavery in 1787, and it subsequently grew into a great national tragedy, the effects of which are still evident. Similarly, without our foresight and action, future generations will attribute the tragedies of climate change and biotic impoverishment to our lies, evasions, and derelictions. But slavery and sustainability also differ in important respects. Slavery was practiced only in a few places, and it could be ended by one means or another. The issues comprising the challenge of sustainability, on the other hand, affect everyone on the Earth for as far into the future as one cares to imagine, and they are for all time. Never again can we take for granted that the planet will recover from human abuse and insult. For all of its complications, slavery was a relatively simple issue compared to the complexities of sustainability. Progress toward sustainability, however defined, will require more complicated judgments involving intergenerational ethics, science, economics, politics, and much else as applied to problems of energy, agriculture, forestry, shelter, urban planning, health, livelihood, security, and the distribution of wealth within and between generations.

Differences aside, Lincoln's example is instructive. He understood that the war had decided only the constitutional issues about the right of states to secede, not the deeper problems of race. It had done nothing to resolve the more volatile problems that created the conflict in the first place. Lincoln had faith that they might someday be solved, but only in a nation in which the better angels of our nature could set aside strife and bitterness. His aim was to create the framework—including the 13th Amendment to the Constitution, which prohibited slavery—in which healing and charity might take root and eventually transform the country.

Lincoln continues to inspire in our time because he framed the legalities of Constitution and war within a larger context of history, obligation, human dignity, and fundamental rights.

The multiple problems of climate change and sustainability will not be solved by this generation, or even the next. Our challenge is to get the name of the thing right and do so in such a way as to create the possibility that they might someday be resolved. Lincoln's example is instructive to us because he understood the importance of preserving the larger framework in which the lesser art of defining particular issues might proceed with adequate deliberation and due process, which is to say that he understood that the art of defining issues is a means to reach larger ends. In our time many things that ought to be and must be sustained are in jeopardy, the most important of which are those qualities Lincoln used in defining the specific matter of slavery: clarity, courage, generosity, kindness, wisdom, and humor.

∞

A second example of leadership instructive to our time is the history of the first one hundred days of Franklin Roosevelt's presidency. In 1933 Roosevelt faced unprecedented challenges of economic collapse and a deteriorating global order. His immediate task was to restore confidence in government, head off total economic ruin, and possibly avert a revolution that many thought to be imminent. The period of his first hundred days is a model for restoring public confidence, even though he did not solve the underlying problems of the economy.

The phrase "the hundred days" was first prominently used to mark the time between Napoleon's escape from Elba and his final defeat at Waterloo in 1815. President Franklin Roosevelt used the phrase to commemorate the period between the opening of the 73rd Congress on March 9, 1933, and its closing on June 17 (Alter, 2006, p. 273; Cohen, 2009). Roosevelt assumed the presidency at the height of the Depression, when the unemployment rate was 25 percent—16 million people were unemployed, and an equal number had only part-time work. The gross national product was half of what it had been four years before, the banking system was on the verge of collapse, the future of democracy in America looked bleak, and fascism was on the march in Europe and the Far East. In those first one hundred days, in historian Arthur Schlesinger's words, Roosevelt:

> sent fifteen messages to Congress, guided fifteen major laws to enactment, delivered ten speeches, held press conferences and cabinet meetings twice a week, conducted talks with foreign heads of state, sponsored an international conference, made all the major decisions in domestic and foreign policy, and never displayed fright or panic and rarely even bad temper. (Schlesinger, 1958, p. 21)

Never before or since has a president displayed a similar energy, such a sure grasp of the realities facing the country, or a deeper understanding of the American people.

Roosevelt aimed first to overcome rampant fear and give people hope, restore confidence in government, and avoid economic collapse. His bearing and personality, honed by the struggle to overcome the effects of crippling polio, were well suited to the challenge. His energy, charm, and political skills were adapted to conditions of a crisis without precedent in U.S. history. He was the first president to travel extensively by airplane and the first to use radio as a tool of mass communication. His approach was more experimental than that of any previous president, and, arguably, more so than any since. In a rare preelection glimpse of his presidency, Roosevelt told a Georgia audience that "The country needs and ... demands bold, persistent experimentation. It is common sense to take a method and try it. If it fails, admit it frankly and try another. But above all, try something." Few took his words seriously. Roosevelt was a complex and paradoxical man, but his politics were pragmatic, not ideological. Yet the measures taken in those first hundred days, however energetic, were not particularly successful. Nor did Roosevelt's New Deal, for all that it accomplished, end the Depression. The Second World War did that. What Roosevelt did do, however, was to restore confidence in the presidency, the government, and particularly the capacity of democracy to confront serious problems.

Barack Obama, the 44th U.S. president, faces challenges that dwarf even those that confronted Lincoln and Roosevelt. At this writing (December 2008) the economy is in free fall, major corporate pillars of the economy are toppling, financial markets have imploded, we are losing two wars, U.S. infrastructure is decrepit, our politics are still bitterly divided, and looming ahead are the multiple challenges of the long emergency. Beyond restoring a semblance of financial order, President Obama must confront for some time to come what Robert Kuttner call "the habits of mind that produced the crisis" (2008, p. 74). He faces as well the large task of recalibrating the office of the presidency to the limits of the Constitution and restoring what political scientist Richard Neustadt once defined as the only real power the president has—the power to persuade. The coercive and manipulative powers of the presidency were enlarged by George W. Bush and Richard Cheney in ways that diminished respect, trust, and effectiveness here and abroad. But unless those enlargements are repudiated by law, all future presidents can—if they choose—wage war preemptively without much interference from Congress, seize and hold American citizens, spy on the citizenry without much if any legal restraint, use practically any federal agency for political purposes, manipulate the press in ways inconceivable prior to 2000, fire federal attorneys for political gain, destroy evidence in criminal cases, use the Justice Department to prosecute members of the opposing party, offer lucrative no-bid government contracts to friends, abet the creation of private security armies, torture, create secret prisons, assassinate inconvenient foreign leaders, circumvent laws by the use of signing statements, and a great deal more. Such things are now possible because the system of checks and balances carefully written into the Constitution and explained in great detail in the *Federalist Papers* was systematically undone, partly as a result of historical circumstances of the 20th century, but with a vengeance by the Bush administration. Said to be necessary in order to protect the country from terrorism, this expansion of presidential authority was in truth carried out by ruthless right-wing ideologues who smelled opportunity in the smoke and ashes of 9/11. While things appeared to be going well, they were abetted by corporate opportunists wanting less regulation and higher profits, the well-to-do wanting lower taxes, a compliant media eager to please, megachurch zealots intending to replace democracy with theocracy, a ragtag army of the congenitally angry, an opposition party that forgot how to oppose, and a drowsy citizenry too distracted to notice the erosion of their liberties. Karl Rove, Dick Cheney, George Bush, and their allies, for a time, conjured up James Madison's worst nightmare—the unification of once carefully separated powers of government—executive, judicial, legislative—in the hands of a single faction, along with substantial control over newspapers, radio, and television and an extensive police and surveillance apparatus he would have loathed. In the words of attorney Scott Horton, "subverting an entire legal apparatus requires great effort. Laws must be circumvented, civil servants thwarted, and opposing politicians intimidated into silence" (2008, p. 38). And they were ... for a time.

President Obama must decide the degree to which he will openly dissociate himself from the expanded powers of the Bush administration. But the historical record gives little encouragement about the contraction of presidential power. Typically, the expanded powers of one president are carefully guarded by successors. The president has distanced himself from the more controversial actions of the Bush administration, but as a matter of political expediency, not for reasons embedded in the Constitution or law. In Horton's words:

we must be prepared to accept a changed system in which the will of the people is subsumed by good manners and fearful politics. As long as this new democracy prevails, little will matter beyond the will of the president." (2008, p. 46)

I do not believe, however, that this is the right, or even a necessary, course. President Obama has other choices, including a thorough investigation of the many irregularities and illegalities involved in the recent dramatic expansion of presidential power. A clear, accurate, and unobstructed record of the Bush years is important, not for purposes of political revenge but to set the record straight—our own version of "truth and reconciliation"—and to restore the office of president to constitutional standards. Many will disagree, saying that learning the truth would be unnecessarily divisive or a waste of time in the face of more pressing business. To the contrary, I believe that we the people, Republicans, Democrats, and independents alike, will need to know the truth in order to reestablish law and order in the highest levels of government and rebuild respect for the office of the president now tarnished by the systematic abuse of power. That restored presidency is, I think, a prerequisite to the transformation necessary to meet the challenges looming ahead. We, the people, will need to know that we are being told the truth and that we are being led by competent, law-abiding, scientifically literate, farsighted, and intellectually engaged public servants who are not beholden to what Theodore Roosevelt once called "malefactors of great wealth" or to any cause beyond the public good broadly conceived. To enter into the long emergency without an accounting for recent presidential abuses is to invite worse in even more difficult times to come.

Looking ahead, it is reasonable to assume that U.S. relations in the Middle East will continue to vex the best minds and subvert our best intentions. Many experts believe that terrorist attacks in the United States aimed at the grid, the Internet, cities, ports, or nuclear power plants are virtually certain. Debt, decaying infra-structure, a national health care emergency, and a badly fractured political system, among other things, will further constrain the choices the president can make, while consuming political atten-tion, energy, and money. But the effects of climate destabilization will soon overshadow every other concern.

The task for the president is to restore trust, rebuild public confidence in government, and provide the leadership necessary to bring order out of our present divisions. In such circumstances, crafting good climate policy by which we might minimize the worst while adapting to what we cannot avoid will be politi-cally difficult but absolutely essential. In journalist and author Tom Friedman's words quoted above, however: "We have not even begun to be serious about the costs, the effort and scale of change that will be required to shift our country, and eventually the world, to a largely emissions-free energy infrastructure over the next 50 years" (2007, p. 42). The long emergency ahead is dif-ferent precisely because it will span virtually every other issue and virtually all aspects of society for a period of time we can scarcely imagine. Climate stabilization and restoration of the biosphere must be made permanent commitments of the nation and must be sustained as a matter of national survival indefinitely.

To that end, in June of 2006, Ray Anderson, Bill Becker, Gary Hart, Adam Lewis, Michael Northrup, and I launched a two-year effort to craft a detailed climate policy for the first hundred days of the administration that would assume office in January of 2009.[1] After dozens of meetings, conference calls, papers, and presentations, the deliberations of several hundred scientists, pol-icy experts, and communications strategists, and presentations to presidential candidates both Republican and Democrat, the final document was handed over to president-elect Obama's transition team in November of 2008. The plan described over 300 pos-sible actions the president could take, as well as a legal analysis of the executive authority at his disposal. Beneath the details, the assumptions that guided the effort were straightforward. The issue of climate destabilization is of such overriding importance that he would have no time for delay and procrastination. The situation called for quick, decisive, effective, and sustained federal action to drastically curtail carbon emissions and deploy renewable energy. On the positive side, an effective energy and climate policy would lessen other problems pertaining to the economy, security, environment, and equity.

It is clear, however, that every president from now on will face many of the same choices, beginning with the question of where to position climate and energy policy in their larger agenda. If this or future presidents regard climate policy as just another problem on a long list of problems, which must therefore compete for resources, funding, and attention with many other issues and the crisis du jour, the chances of failure on all counts will be greater. If a climate and energy policy, however, are permanently regarded as the linchpin connecting other issues, including policies for security, economy, environment, and justice, the road ahead will be a great deal easier, and the chances of coming through the long emergency with the nation intact will be much higher.

The immediate situation that President Obama faced was rendered more difficult because capacity and morale in many government departments and agencies had been badly eroded in prior years. The remedy requires both engaging and retaining talented and committed people in public service and the adroit remodeling of the machinery of government. In addition to selecting members of the cabinet, President Obama made another 7,000 or so appointments to positions in the federal government, including 400–500 members of the White House staff and 1,200–1,300 in the executive office of the president. As always, the ability to implement policy of any kind depends in large part on the intellect, experience, energy, creativity, character, and personal skills of presidential appointees and White House staff. Beyond administrative and executive skills, from now on those in such positions must also understand ecology, Earth systems science, and the multiple ways in which public policy and natural systems interact—what is emerging as "the new science of sustainability" (Goerner, Dyck, and Lagerroos, 2008).

In the years ahead, the situation will likely get a great deal worse before it improves. This president and those to follow, accordingly, must communicate in ways that sustain public morale and keep the vision of a sustainable society in clear focus. In his inaugural address, Roosevelt, the master psychologist, aimed to calm public fears: "The only thing we have to fear is fear itself." But that was a public on the edge of desperation. The public presently may or may not be less fearful, but it is certainly more confused about climate change and what can be done about it at a time of economic distress. It is perhaps beyond the "tipping point" of awareness about the issue, but has not yet grasped the seriousness of climate change or the kind of choices that must be made. As the effects of climate destabilization become more apparent, however, public apathy and confusion may shift to desperation, panic, and possibly the search for scapegoats. In those circumstances, presidents can choose communications strategies that range from Churchill's "blood, toil, tears, and sweat" approach to sunny optimism at the other extreme. In either case, as the situation becomes darker, presidents must appeal to the better angels of our nature, framing the issues as those of intergenerational fairness and morality. Whatever the specific content, the president's communication strategy, like Lincoln's, ought to rise above the divisions of right and left, liberal and conservative, to identify common interests and present a vision of higher ground beyond.

The substance and style of political communication must be matched, in other words, to the time and to the public need for reassurance, clear direction, and honest information artfully delivered. In recent decades, however, the standards for presidential communication have fallen considerably, victim of the demands of television, which emphasize appearance over content, and polling that exaggerates short-term political gains over long-term public realities, as well as the tawdry politics of a abnormally corrupt era. But we have better models. Teddy Roosevelt used the presidency as a "bully pulpit." His cousin Franklin in dark times used radio, creatively adapted for "fireside chats," with extraordinary results. John F. Kennedy was a master of the art of press conferences. In the presidential campaign of 2008, Barack Obama used the Internet adroitly to reach the young and new voters. In the Internet era, the tools of communication have multiplied many times over. But whatever the medium—television, radio, press conferences, personal appearances, Internet, or public addresses—presidents must craft communication strategies that lift the public out of apathy or despair while educating, informing, and inspiring by showing a plausible way forward consonant with our obligations, our national heritage, and global realities. Presidents must lay the groundwork for a durable and broad coalition around the national interest in climate stability that protects our long-term security and distributes the costs and benefits fairly within and between generations.

All such efforts could come to naught, however, if access to the public airwaves is not restored to public control. Public confusion and ignorance about energy issues and climate science plays to the advantage of the fossil fuel industry, and if allowed to continue will sharply diminish our prospects in the years ahead. Co-optation of the public airwaves by corporate interests can drown out the voice of the public as well as that of even the most eloquent president. To ensure that the public is adequately informed, not misled and deliberately confused, the president, again, should direct the Federal Communications Commission, among other things, to reinstate the "fair and balanced" standard as a requirement for the use of the public airwaves. The integrity of broadcasting is essential to educate the public about the choices ahead and their consequences.

The fact is that we are now making the most fateful policy decisions that humans will ever make. The choice of tools by which those decisions will be implemented range from free-market approaches to use of the police power of the state to mandate changes. Some propose raising taxes on energy, while others advocate placing a cap on carbon emissions and allowing emitters to buy and sell permits. How we decide and what we decide will greatly affect our prospects in the long emergency. And whatever the specific policy tools selected they must be flexible enough to be made more stringent as evidence warrants.

Broadly, we must choose between energy policies that emphasize efficiency, renewable energy, and better design that eliminates much of the need for energy in the first place (Kutscher, 2007; Makhijani, 2007) and "hard," expensive, and large-scale options such as continued use of coal with carbon sequestration and nuclear power. The choices and their relative consequences must be made crystal clear to the public despite the well-funded efforts by the coal industry to promote the fiction of "clean coal" and equally well-funded efforts by ardent revivalists to resuscitate nuclear power. We will have neither the time nor the money to undo expensive mistakes later. Accordingly, the president must set the framework for a rational public dialogue about energy policy in which all options are compared on a level playing field that includes criteria such as:

1. Carbon eliminated per dollar spent;
2. Energy return on investment;
3. The speed with which technology can be deployed;
4. Near-term technical feasibility; and
5. Resilience in the face of malfeasance, acts of God, and human error.

And good policy will not simply switch problems but will solve them while protecting public safety and health.

The president has the power to define the larger political topography on which climate policy is debated. It is possible to craft policies that join conservatives and liberals in ways that are transparent, pragmatic, and fair. The outlines of consensus have emerged around an energy policy that would:

Reduce our dependence on imported fuels
Minimize our vulnerability to political conflicts in unstable parts of the world
Reduce our balance of payments deficit
Raise the costs of fossil fuels relative to those of improved efficiency and renewables
Lower costs that are now externalized
Generate better technology and a stronger economy
Create millions of jobs in green energy
Improve air and water quality
Protect public health
Lower health costs
Reduce the influence of entrenched energy industries on U.S. politics.

The right energy policy, in other words, should solve or lessen many problems, including those of climate change and national security, while providing many collateral benefits. The president will need to sell such a vision in order to build a majority coalition that crosses old lines of left and right by joining people of faith, laborers, farmers, minorities, business leaders, intellectuals, and members of the financial community.

Whatever the content of the policy and however it is communicated, effective implementation will require changes at the interface between policy and science at all levels. The first objective is to rebuild and enhance the integrity and capacity of federal environment and science-based agencies (such as the Department of the Interior, the Environmental Protection Agency, the Department of Energy, the National Aeronautics and Space Administration, and the National Oceanic and Atmospheric Administration) that are essential to climate research and our capacity to foresee

and forestall. Second, in order to consistently implement climate policy across all branches of government, this and all future
presidents will need the capacity to coordinate actions of federal
departments and agencies that now often conflict. The president
and Congress will need an expanded federal capacity to assess
technologies similar to that once provided by the Congressional
Office of Technology Assessment, which was terminated in Newt
Gingrich's "Contract with America" in 1994. In addition, we will
need to expand the foresight capacity of government in novel
and creative ways that more fully engage the National Academy
of Sciences; the broader scientific community represented by the
American Association for the Advancement of Science; federal
laboratories, such as the National Renewable Energy Laboratory; and the president's science advisor. We need new ideas as
well. British journalist George Monbiot, for example, proposes a
"100-year committee" whose purpose would be "to assess the
likely impacts of current policy in 10, 20, 50, and 100 years' time."
Governments, by whatever means, must learn to reconcile short-
term imperatives with long-term trends in imaginative and effective ways. But that requires that those presuming to govern have
the capacity and willingness to see connections across political
lines, geography, species, and time.

In the decades ahead, presidents will also need a greater capacity to respond quickly and effectively to climate-driven disasters.
The failure of the federal response to the devastation caused by
Hurricane Katrina, again, is a textbook example of what not to
do. We must anticipate and prepare for a future in which hurricanes, large storms, flood, fire, drought, and acts of terrorism may
become the norm. For that reason the capacity of federal government to respond to emergencies must be much more robust
and effective, not just for occasional events but for multiple and
perhaps frequent events.

But no such changes are likely until the great encampment
of K Street lobbyists is disbanded and sent packing. They have

corrupted democracy and undermined our prospects for too long. Despite recent and voluminous evidence of influence peddling and scandal, their power is scarcely diminished and poses a significant threat to any effective climate and energy policy. President Obama and those to follow must permanently curtail the power of money in U.S. politics. While there is no shortage of ideas on how to do so, the best solution would be to remove money from the electoral process once and for all by publicly financing elections to national offices.

Early on, federal policy must encourage climate mitigation and adaptation at state and local levels. Presidential leadership will be necessary to cement partnerships with governors, mayors, and business leadership to build locally and regionally resilient economies, food systems, and distributed energy networks that would enhance the capacity to withstand the disruptions of climate change. Distributed energy in the form of widely dispersed solar and wind technology would buffer communities from supply interruptions, failure of the electrical grid, and the shock of sudden price increases. Similarly, the resuscitation of local agriculture would reduce dependence on long-distance transport from distant suppliers.

In the 1930s, President Roosevelt experimented with a variety of ways to put Americans to work doing useful things. The Civilian Conservation Corps, for example, put the unemployed and youth to work building roads, schools, and public buildings and restoring the public lands. Updated to the 21st century, that model is a good one for the establishment of green economies such as that proposed by attorney Van Jones to engage the disadvantaged and unemployed in a new green economy built from the bottom up (Jones, 2008). The energy and creativity of young people, trained in renewable energy technologies, could be deployed to build wind farms, install solar technologies, and improve energy efficiency in low-income communities while creating millions of new jobs.

Not least, the 44th president and all his successors must bring America back into the international community on climate, security, and economic issues. There is no prospect of stabilizing climate without a coordinated global effort that systematically addresses carbon emissions, poverty, and security (Hart, 2006; Speth, 2004 and 2008). The world is waiting for U.S. leadership to help create a global partnership on climate policy. Given the disparity of historic and present carbon emissions, the United States is obliged to set the right example and take the lead. But no one should assume that the present hostility toward the United States will disappear without a significant effort sustained over many years.

Presidential leadership has many intangibles. The president has the power to issue executive orders that affect government purchasing and management of federal facilities, among other things. Presidents have the power to initiate change in both statutory and regulatory law. But in the final analysis, this and future presidents must use all of their persuasive powers to encourage the American people to move resolutely, boldly, and quickly toward a much better future than that in prospect. If this opportunity is lost, there will likely be no other.

This president and his successors will have a great deal of persuading to do. Others could assume a continually growing economy and thereby avoid thorny issues about fair income distribution by hiding behind old myths that rising tides that would lift all boats or wealth would trickle down from the tables of the rich. The prospects for continual economic growth, however, are not good. Even Nobel Prize–winning economist Robert Solow now admits to being agnostic on the possibilities of continued growth. On a finite planet governed by the laws of thermodynamics and ecology, the reasons are not difficult to find. Economists of the neoclassical persuasion assumed that economic growth would be possible by substituting more abundant for scarce resources and by increasingly heroic technology. The results, however, are rapidly diminishing resources along with rising inequity, ecosystems verging on collapse, and rapid climate change.

There are many reasons why an economic doctrine over two centuries old and on speed for the last 60 years could not work for long in a biosphere with rules evolved over 3.5 billion years. They are well described by economists such as Herman Daly, Robert Costanza, Peter Victor, and Richard Douthwaite. The point is that this president and those to follow must begin the difficult but necessary task of educating the public to understand why the growth economy measured as quantity will end one way or another and must be replaced by an economy based on durability, quality, and fairness. It will be necessary to state the obvious but often overlooked facts that the benefits of the growth economy were distributed unfairly, and often in ways that were ugly and inconvenient to boot. Urban sprawl, for example, fueled a great deal of economic growth, but also squandered energy, time, land, people, and quality of life. For those left behind in the cities, the experience of growth has often been a nightmare of crime, diminished services, bad schools, and joblessness. In either case, beyond some fairly minimal point, the growth economy did not contribute nearly as much to our well-being and happiness as promised and is not sustainable anyway. The challenge for this and future presidents is to use their authority and powers to free us from the grasp of outworn economic doctrines and align our expectations and behavior with biophysical realities.

∽

True leadership, such as that shown by Lincoln and Franklin Roosevelt, is the rarest of human traits. It is often confused with management, which keeps the trains running on time and paper moving efficiently through the bureaucracy.[2] Management is about efficiency, what German sociologist Karl Mannheim (1940) once called "functional rationality," while leadership has to do with "substantive rationality," or the big choices that we make. Leadership is about direction, relating our highest ideals to our

actions. It is first and foremost about direction, and only second-
arily about the "framing" of issues. True leaders do not rely on
what has lately been called "spin" or manipulation as much as
on truth telling and the arts of rational argument expressed with
eloquence and passion. Real leaders help others see the topography of history through the fog of events and crises. Real leaders
energize people to act better than they otherwise would. Some
are charismatic visionaries, but more often they are rather ordinary persons, blessed with a capacity to ask good questions, see
patterns, notice when the emperor is naked, and act while others sit quietly. As Robert Coles and Garry Wills, among others,
note, leadership comes in a variety of forms and scales, in many
areas of life. And, of course, history, including our own, is full of
examples of people in positions of potential leadership who led in
the wrong direction, for which we are paying a high price.

Many people dismiss leadership as mostly pathological and
increasingly irrelevant. The real driver of progressive social change
supposedly comes from the bottom of society and works upward.
They envision a networked world in which electronic spontane-
ity and the basic wisdom of the human heart displace the need
for leaders at all. Something is astir in the world, abetted by the
ease of e-mail, the Internet, and cell phones. Whether this will
amount to much I cannot say. Social movements fail as often as
not because their goals are inchoate, they work at cross-purposes,
or they fall victim to one human fault or another, or because
circumstances and the tide of history work against them. The
Populist movement in the late 19th century, for example, had a
great deal going for it, but it failed in large part because of the
power of capital and because rural populists and the urban labor
movement could not come together around a common agenda.
Similarly, what is presumed to be a rising tide of contemporary
social movements has yet to show that it has enough common
purpose, unity, and vitality to cause change of the right sort in
the time available.

But the dichotomy of leaders versus mobilized publics is a false one; both are necessary, either alone is not enough. We will need both because the relation between them is reciprocal. No would-be leader anywhere can get too far out in front of the public. "An activated citizenry," in Robert Kuttner's words, "is not just a passive army of political supporters who will cheer and vote on cue" (2008, p. 109). Movements, however, can be formless and ineffective without the focusing effect of visionary leadership. It would be difficult to imagine the movement to liberate India without Gandhi, or Gandhi without the force of the people behind him. We do not yet know whether the Internet will amplify and accelerate progressive social movements more than those pushing in the other direction. Neither do we know whether advancing communications technology will be immune to various corruptions characteristic of other forms of communication.

We do know, however, that addressing climate destabilization will require presidential leadership of a high order *and* a massive amount of what Paul Hawken calls "blessed unrest," and lots of other things as well. But presidential leadership on any issue of human survival has been rare. One exception was John F. Kennedy's speech at American University in June of 1963, which led to the Nuclear Test Ban Treaty. Another was Ronald Reagan's attempt with Mikhail Gorbachev at the Reykjavík Summit in October 1986 to abolish nuclear weapons. By some accounts, they came close. But on issues of sustainability in general and climate change in particular, no president until Obama has yet offered vision and leadership, or used the White House as a "bully pulpit" in the manner of Theodore Roosevelt. Instead, presidents have mostly ignored, evaded, and, most recently, actively denied the reality of the issue. If we are to avoid the worst that could happen, however, the United States must rejoin the world community in the effort to stabilize and reduce greenhouse gases, and that will require, in turn, active, creative, and transformational leadership by this president and all thereafter.

PART II

CONNECTIONS

CHAPTER 4

The Carbon Connection

Combustion is the hidden principle behind every artefact we create... From the earliest times, human civilization has been no more than a strange luminescence growing more intense by the hour, of which no one can say when it will begin to wane and when it will fade away. For the time being, our cities still shine through the night, and the fires still spread.

—W. G. Sebald

Civilization is an experiment, a very recent way of life in the human career, and it has a habit of walking into what I am calling progress traps... The most compelling reason for reforming our system is that the system is in no one's interest. It is a suicide machine.

—Ronald Wright

HAVING SEEN PICTURES OF THE DEVASTATION DID not prepare me for the reality of New Orleans. Mile after mile of wrecked houses, demolished cars, piles of debris, twisted and downed trees, and dried mud everywhere. We stopped every so often to look into abandoned houses in the 9th Ward and along the shore of Lake Pontchartrain to see things close up: mud lines on the walls, overturned furniture, moldy clothes still hanging in closets, broken toys, a lens from a pair of glasses... once cherished and useful objects rendered into junk. Each house had a red circle painted on the front to indicate the results of the search for bodies. Some houses showed the signs of desperation, such as holes punched through ceilings as people tried to escape rising water. The musty smell of decay was everywhere, overlaid with an oily stench. Despair hung like Spanish moss in the hot, dank July air.

Ninety miles to the south, the Louisiana delta is rapidly sinking below the rising waters of the Gulf. This is no "natural" process but rather the result of decades of mismanagement of the lower Mississippi, which became federal policy after the great flood of 1927. Sediments that built the richest and most fecund wetlands in the world are now deposited off the continental shelf—part of an ill-conceived effort to tame the river. The result is that the remaining wetlands, starved for sediment, are both eroding and compacting, sinking below the water and perilously close to no return. Oil extraction has done most of the rest of the damage by crisscrossing the marshlands with channels that allow the intrusion of saltwater and storm surges. Wakes from boats have widened the original channels considerably, further unraveling the ecology of the region. The richest fishery in North America and a unique culture that once thrived in the delta are disappearing, and with them the buffer zone that protects New Orleans from hurricanes. "Every 2.7 miles of marsh grass," in Mike Tidwell's words, "absorbs a foot of a hurricane's storm surge" (2003, p. 57).

And the big hurricanes will come. Kerry Emanuel, an MIT scientist and former climate change skeptic, researched the connection among rising levels of greenhouse gases in the atmosphere, warmer sea temperatures, and the severity of storms. He's a skeptic no longer (2005, pp. 686–88; also Trenberth, 2007). The hard evidence on this and other parts of climate science has moved beyond the point of legitimate dispute. Carbon dioxide, the prime greenhouse gas, is at the highest level in at least the last 650,000 years and probably a great deal longer, and it continues to accumulate by ~2.0+ parts per million per year, edging closer and closer to what some scientists believe is the threshold of runaway climate change. British scientist James Lovelock compares our situation to being on a boat upstream from Niagara Falls with the engines about to fail.

If this were not enough, the evidence now shows a strong likelihood that sea levels will rise more rapidly than previously thought. The Third Assessment Report of the Intergovernmental Panel on Climate Change in 2001 predicted less than a one-meter rise in the 21st century, but more recent estimates put this figure higher as a result of accelerated melting of the Greenland ice sheet and polar ice along with the thermal expansion of water (Overpeck et al., 2006; U.S. Climate Change Science Program, 2009; Smith et al., 2009; McKie, 2009).

Nine hundred miles to the northeast of New Orleans as a sober crow would fly it, Massey Energy, Arch Coal, and other companies are busy leveling the mountains of Appalachia to get at the upper seams of coal in what was one of the least disturbed forests in the United States and one of the most diverse ecosystems anywhere. Throughout the coalfields of West Virginia and Kentucky they have already leveled 500 mountains, give or take a few, across 1.5 million acres, and they intend to destroy a good bit more. These companies wash coal on-site, leaving billions of gallons of a dilute asphalt-like gruel laced with toxic flocculants and heavy metals. An estimated 225 such containment ponds are located over abandoned mines in West Virginia, held back from the communities below only by earthen dams prone to failure either by collapse or by draining down through the old mine tunnels that honeycomb the region. One of these dams failed on October 11, 2000, in Martin County, Kentucky, when the slurry broke through a thin layer of shale into mines and then into hundreds of miles of streams and rivers. The result was the permanent destruction of waterways and of the property values of people living in the wake of an ongoing and mostly ignored disaster. This is typical of the coalfields. They are a third-world colony within the United States, a national sacrifice zone in which fairness, decency, and the rights of old and young alike are discarded as unnecessary on behalf of the national obsession with "cheap" electricity.

Jack Spadaro is a heavyset, rumpled, and bearded man with the knack for describing outrageous things calmly and with clinical precision. A mining engineer by profession, he spent several frustrating decades trying to enforce the laws, such as they are, against an industry with friends in high places in Charleston, Congress, and the White House. For his role in trying to enforce even the flimsy laws that might have held Massey Energy slightly accountable for its flagrant and frequent malfeasances, the Bush administration tried unsuccessfully to fire Jack from his position as a mine safety inspector in the Interior Department but eventually forced him to retire.

He is in the first plane to take off from Yeager Field in Charleston, along with the chief attorney for Wal-Mart, the largest corporation in the world. Hume Davenport, founder of SouthWings, Inc., is the pilot of the four-seat Cessna. The ground recedes below us as we pass over Charleston and the Kanawha River, lined with barges hauling coal to power plants along the Ohio River and points more distant. Quickly appearing on the western horizon is the John Amos plant owned by American Electric Power. We are told that the plant releases more mercury to the environment than any other facility in the United States, as well as hundreds of tons of sulfur oxides, hydrogen sulfide, and CO_2 each year. For a few minutes we can see the deep green of wrinkled Appalachian hills below, but very soon the first of the mountaintop removal sites appears. It is followed by another and then another. The pattern of ruin spreads out below us for miles, stretching to the far horizon on all points of the compass. From a mile above, trucks with 12-foot-diameter tires and drag lines that could pick up two Greyhound buses at a single bite look like Tonka toys in a sandbox. What is left of Kayford Mountain comes into sight. It is surrounded by leveled mountains and a few still being leveled. "Overburden," the mining industry term for dismantled mountains, is dumped into valleys covering hundreds of miles of streams—an estimated 1,500 miles in the past 25 years. Many more miles of streams will be buried if the coal companies have their way. Coal slurry ponds loom above houses, churches, and even elementary schools. When the earthen dams break on some dark rainy night, those below will have little if any warning before the deluge hits.

Jack is our guide to the devastation. In a flat, unemotional monotone he explains what we are seeing below. Aside from the destruction of the Appalachian forest, the math of the reengineered mountain remnants, he explains, is all wrong. The slopes are too steep, the slurry impoundments too large. The angles of slope, the weight against the dams, and the proximity of houses and towns are the geometry of tragedies to come. Spadaro points out the Marsh Fork Elementary School situated close to a coal loading operation and below a huge impoundment back up in the hollow. In the event of a dam failure, the evacuation plan calls for the principal, using a bullhorn, to sound the alarm and begin the evacuation of the children ahead of the 50-foot wall of slurry that will be moving toward them at upwards of 60 miles an hour. If all works according to the official evacuation plan, they will have two minutes to get to safety, but once out of the school there is no safe place for them to go.

So it is in the coalfields—ruin at a scale for which there are no adequate words; ecological devastation to the far horizon of topography and time. We say that we are fighting for democracy elsewhere, but no one in Washington or Charleston seems aware that we long ago deprived some of our own the rights to life, liberty, and property.

On the circle back to Yeager field in Charleston, Tom Hyde, the Wal-Mart attorney, calls the devastation "tragic." We all nod, knowing the word does not quite describe the enormity of the things we've just seen or the cold-blooded nature of it. In our one-hour flight we saw maybe 1 percent of the destruction now metastasizing through four states. Until recently it was all but ignored by the national media. We have known of the costs of mining at least since Harry Caudill published *Night Comes to the Cumberlands* in 1963, but we have yet to summon the moral energy to resolve the problem or pay the full costs of the allegedly cheap electricity that we use.

Under the hot afternoon sun we board a 15-passenger van to drive out to the edge of the coalfields to see what they look like on the ground. We take the interstate south from Charleston and exit at a place called Sharon onto winding roads that lead to mining country. Trailer parks, small evangelical churches, truck-repair shops, and small, often lovingly tended houses line the road, intermixed with those abandoned long ago when underground mining jobs disappeared. The two-lane paved road turns to gravel and climbs toward the top of the hollow and Kayford Mountain.

Within a mile or two, the first valley fill appears. It is a green V-shaped insertion between wooded hills. Reading the signs made by water coursing down its face, Spadaro notes that this one will soon fail. Valley fills are mountains turned upside down: rocky mining debris, illegally burying trees, along with what many locals believe to be more sinister things brought in by unmarked trucks in the dead of night. He adds that some valley fills may contain as much as 500 million tons of blasted mountains and run for as long as six miles. We ascend the slope toward Kayford, passing by the "no trespassing" signs that appear around the gate that leads to the mining operations.

Larry Gibson, a diminutive bulldog of a man fighting for his land, meets us at the summit, really a small peak on what was once a long ridge. His family has operated a small coal mine on Kayford since 1792. Larry, the last of the Kayford Gibsons, is the proverbial David fighting the Goliath of Massey Energy, but he has no slingshot other than that of moral authority spoken with a fierce, inborn eloquence. Those traits, and the raw courage he shows every day, have made Larry a poster child for the movement to stop mountaintop removal, with his story told in *Vanity Fair,* in *National Geographic,* and on CNN. Larry's land has been saved so far because he made 40 acres of it into a park and has fought tooth and nail to save it from the lords of Massey Energy. They have leveled nearly everything around him and have punched holes underneath Kayford because the mineral rights below ground and the ownership of the surface were long ago separated in a shame-less scam perpetrated on illiterate and trusting mountain people who had ousted earlier illiterate and trusting inhabitants.

Larry describes what has happened using a model of the area that comes apart more or less in the same way that the mountains around him have been dismantled. As he talks he illustrates how mountains are taken down by taking the model apart piece by piece, leaving the top of Kayford as a knob sticking up amidst the encircling devastation. So warned, we walk down the country lane to witness the advancing ruin. A mother bear with her cubs was said to have run down this road the day before, fleeing the dev-astation. Fifteen of us stand for maybe a half an hour on the edge of the abyss watching giant bulldozers and trucks at work below us. Plumes of dust from the operations rise up several thousand feet. The next set of explosive charges is ready to go on an area below us that appears to be about the size of a football field. Every day some three million pounds of explosives are used in the 11 counties south of Charleston. This is a war zone. The mountains are the enemy, profits from coal the prize, and the local residents and all those who might have otherwise lived here or would have vacationed here are the collateral damage.

We try to wrap our minds around what we are seeing, but words do no justice to the enormity of it. Some of the oldest mountains on Earth are being turned into gravel for a pittance; their ecologies radically simplified, forever. Perhaps as a defense mechanism from feeling too much or being overwhelmed by what we've seen, we talk about lesser things. On the late after-noon drive back to Charleston, we pass by the coal-loading facili-ties along the Kanawha River. Mile after mile of barges are lined up to haul coal to hungry Ohio River power plants, the umbilical cord between mountains, mines, and us—the consumers of cheap electricity.

Over dinner that night we hear from two Mingo County resi-dents who describe what it is like to live in the coalfields. Without forests to absorb rainwater, flash floods are a normal occurrence. A three-inch rain can become a ten-foot wall of water cascading off the flattened mountains and down the hollows. The mining industry calls these "acts of God" and public officials—thoroughly bought or intimidated or both—agree, leaving the victims with no recourse. Groundwater is contaminated by coal slurry and the chemicals used to make coal suitable for utilities. Well water is so acidic that it dissolves pipes and plumbing fixtures. Cancer rates are off the charts, but few officials in Charleston or Washington notice. Coal companies are major buyers of politicians, and the head of Massey Energy, Donald Blankenship, has been known to spend lots of money to buy precisely the kind of representatives he likes—the sort who can accommodate themselves to exploitation of land and people and the profits to be made from it. His campaign to ravage the rest of West Virginia has the Rovian title "For the Sake of the Kids."

Pauline and Carol from the town of Sylvester, both in their seventies, are known as the "dust busters" because they go around the town wiping down flat surfaces with white cloths that are covered with coal dust from a nearby loading facility. These are presented at open hearings as evidence of foul air to the irritated and unmovable servants of the people. Black lung and silicosis disease are now common among young and old alike who are exposed to dust from surface operations but who have never set foot in a mine. They have little or no voice in government; they are considered to be expendable. Pauline, a fiercely eloquent woman, whose husband was wounded and captured by the Germans in the Battle of the Bulge in 1944, rhetorically asks, "Is this what he fought for?" The clock reads 9:30 P.M., and we quit for the day.

To permanently destroy millions of acres of Appalachia in order to extract maybe 20 years' worth of coal representing perhaps 3 percent of our national coal use is a form of insanity at a scale that I cannot adequately describe. As a nation we have so far lacked the compassion and good sense to stop it, and all the talk about "clean coal" notwithstanding, there is no decent case to be made for it. Unlike deep mining, mountaintop removal employs few workers. It is destroying the wonders of the mixed mesophytic forest of northern Appalachia, including habitat for dozens of endangered species, once and for all. It contaminates groundwater with toxics and heavy metals and renders the land permanently uninhabitable and unusable. Glib talk of the economic potential of flatter places for commerce of one kind or another is just that: glib talk. Coal companies' efforts to plant grass and a few trees here and there are like putting lipstick on a corpse. The fact of the matter is that mountaintop removal is destroying one of the most diverse and beautiful ecosystems in the world, rendering it uninhabitable forever. And it is destroying the lives and culture of the people who have stayed behind in places like Sylvester and Kayford.

Some politicians, energy corporation executives, and experts justify the devastation on the grounds of necessity and cost. But virtually every independent study of energy use done in the past 30 years has concluded that we could cost-effectively eliminate half or more of our energy use while strengthening our economy, reducing the incidence of asthma and lung disease, raising our standard of living, and improving environmental quality. A more complete accounting of the benefits of reducing coal use would also include avoidance of the inevitable tide of damage and insurance claims attributable to climate change. Some say that if we don't burn coal, the economy will collapse and we will all have to go back to the caves. But with wind and solar power growing by 40 percent or more per year and the technology of energy efficiency advancing rapidly, we have good options that make burning coal unnecessary. And before long we will wish that we had not destroyed so much of the capacity of the Appalachian forests and soils to absorb the carbon that makes for bigger storms and more severe heat waves and droughts.

Very few in positions of authority in West Virginia politics, excepting that noble patriarch of good sense, Ken Hechler, ask the

obvious questions. How far does the plume of heavy metals coming from coal-washing operations go down the Kanawha, Ohio,
and Mississippi rivers and into the drinking water of communities elsewhere? What other economy—based on wind power,
land restoration, the sustainable use of forests, nontimber forest
products, ecotourism, and human craft skills—might still flourish
in the remnants of these ancient hills? What is the true cost of
"cheap" coal? Why do the profits from coal mining always leave
the state? Why is so much of the land owned by absentee corporations? Why do so few hold absolute power over so many and
so much?

Once you subtract the permanent ecological ruin and crimes
against humanity, there really isn't much to add, as a country song
once put it. Believers in "clean coal" ought to spend some time in
the coalfields, survey the ruined hills and lives, and talk to the residents in order to understand what those words really mean at the
point of extraction. And for all of the talk about safely and permanently sequestering carbon from burning coal, there is precious
little evidence that it could be done at all, or were it possible, that
it could compete with improved energy efficiency and renewable energy. "Clean coal" is a scam foisted on the gullible by the
coal companies hoping for a few more years of profit at a cost we
cannot fathom.

Nearly a thousand miles separate the coalfields of West Virginia
from the city of New Orleans and the Gulf Coast, yet in some
important ways they are a lot closer than can be measured in miles.
The connection is carbon. Coal is mostly carbon, and for every
ton of coal burned, 3.6 tons of CO_2 eventually enter the atmosphere and remain there for one or two centuries, raising global
temperatures, warming oceans and thereby creating bigger storms,
melting ice, and raising sea levels for a long time to come. And
between the remaining hills of Appalachia and the sinking land
of the Louisiana coast, tens of thousands of people living downwind from coal-fired power plants die prematurely each year from

breathing small particles in the smoke that are laced with heavy metals and that penetrate deeply into lung tissue.

While the issues and solutions are complex, the underlying problem is very simple. It is a dance of mutual ruin. Some of the oil extracted in the Gulf of Mexico powers the pickup trucks and mining equipment in coalfields as well as the trains that haul coal to power plants. The carbon emitted from those power plants eventually amplifies the storms and sea-level rise that will doom the oil industry, if scarcity and economic turmoil don't get it first. And the oil extraction business has helped to destroy coastal ecologies that buffer the land from larger storms.

✺

In 1987 the World Commission on Environment and Development launched a global debate about how to make economic development sustainable. Not surprisingly, the language and recommendations in the final report were crafted to appeal to everyone—bankers and environmentalists, CEOs and citizens everywhere alike. Its message was that the "present generation could meet its needs without depriving the future," which is to say that if we are only a little smarter, all can go on as before. The authors aimed to avoid giving offense and challenging existing priorities, so they mostly ignored unpleasant things like the limits of the Earth's carrying capacity, fair distribution of risks, costs, and benefits, and the need to reconcile infinite human demands with the limits of a finite planet. They proposed to fine-tune the growth juggernaut, but nothing more discomfiting.

Nearly a quarter of a century later, however, world population has grown by nearly two billion, the gross world product has doubled, energy use has grown by 42 percent, water use is reaching critical proportions, 90 percent of the large fish in the oceans are gone, the climate is trending toward destabilization, and the disparity between the richest and poorest continues to widen. By one estimate, the human footprint exceeds the capacity of the Earth by 25 percent, and the deficit of carrying capacity continues to grow. Were everyone to live like Americans, humankind would require the resources of three additional Earths. After a quarter of a century of "sustainable development," virtually no indicator of planetary health is moving in a positive direction, and we should ask why (Speth, 2008).

It is clear by now that we have seriously underestimated the magnitude and speed of the human destruction of nature, but we seem powerless to stop it. The rapid destabilization of climate and the destruction of the web of life are just symptoms of larger issues the understanding of which runs hard against our national psyche and the Western worldview generally. Americans, the largest users of Prozac, proudly think of themselves as an optimistic, "can-do" people not easily given to doubt or despair. That is mostly a useful outlook, until it is not. According to Evan Connell, George Armstrong Custer's last recorded words just before the opening shots at the battle of Little Big Horn were a stirring: "Hurrah boys, now we have them" (Connell, 1985, p. 279). Optimistic bravado in the face of long odds is a much-admired American trait in some quarters, and sometimes it works out, sometimes it doesn't. Unfortunately for Custer, that day Sitting Bull and the Sioux were not much amused and apparently not particularly awed by the chutzpah of the 7th Cavalry. In our own time, the pronouncement by George Herbert Walker Bush that "The American way of life is not negotiable" came only a decade before Osama bin Laden negotiated it downward several trillion dollars, depending on how much you care to include. The end of cheap oil will take it down several more notches.

It is easier, I think, to understand the reality of dilemmas in places that have historic ruins and are overlaid with memories of tragedies and misfortunes that testify to human fallibility, ignorance, arrogance, pride, overreach, and sometimes evil. Amidst shopping malls, bustling freeways, and all of the accoutrements, paraphernalia, enticements, and gadgetry of a booming fantasy industry, it is harder to believe that sometimes things don't work out because they simply cannot or that limits to desire and ambition might really exist. When we hit roadblocks, we have a national tendency to blame the victim or bad luck, but seldom the nature of the situation or our beliefs about it. What Spanish philosopher Miguel de Unamuno (1977) called "the tragic sense of life" has little traction just yet in the United States, because it runs against the national character and we don't read much philosophy anyway.

A tragic view of life is decidedly not long-faced and resigned, but neither is it giddy about our possibilities. It is merely a sober view of things, freed from the delusion that humans should be about "the effecting of all things possible" or that science should put nature on the rack and torture secrets out of her, as we learned from Francis Bacon. It is a philosophy that does not assume that the world or people are merely machines or that minds and bodies are separate things, as we learned from Descartes. It is not rooted in the assumption that what can't be counted does not count, as Galileo argued. The tragic sense of life does not assume that we are separate atoms, bundles of individual desires, unrelated, hence without obligation to others or what went before or those yet to be born. Neither does it assume that the purpose of life is to become as rich as possible for doing as little as possible, or that being happy is synonymous with having fun. The tragic view of life, on the contrary, recognizes connections, honors mystery, acknowledges our ignorance, has a clear-eyed view of the depths and heights of human nature, knows that life is riddled with irony and paradox, and takes our plight seriously enough to laugh at it.

Whether aware of it or not, all of us are imprinted with the stamp of Bacon and the others who shaped the modern worldview. The problem, however, is not that they were wrong but rather that we believed them too much for too long. Taken too far and applied beyond their legitimate domain, their ideas are beginning to crumble under the weight of history and the burden of a reality far more complex and wonder-filled than they knew or could have known. Anthropogenic climate destabilization is a symptom of something more akin to a cultural pathology. Dig deep enough and the "problem" of climate is not reducible to the standard categories of technology and economics. It is not merely a problem awaiting solution by one technological fix or another. It is, rather, embedded in a larger matrix, a symptom of something deeper. Were we to "solve" the "problem" of climate change, our manner of thinking and being in the world would bring down other curses and nightmares now waiting in the wings. Perhaps it would be a nuclear holocaust, or terrorism, or a super plague, or, as Sun Microsystems founder Bill Joy warns, an invasion of self-replicating devices like the products of nanotechnology, genetically engineered organisms, or machines grown smarter than us that will find us exceedingly inconvenient.[1] There is no shortage of such plausible nightmares, and each is yet another symptom of a fault line so deep that we hesitate to call it by its right name.

The tragic sense of life accepts our mortality, acknowledges that we cannot have it all, and is neither surprised nor dismayed by human evil. The Greeks who first developed the dramatic art of tragedy knew that we are ennobled not by our triumphs or successes but by rising above failure and tragedy. Sophocles, for example, portrays Oedipus as a master of the world—powerful, honored, and quite full of himself, but also honest enough to search out the truth relentlessly, and in doing so he falls from the heights. That is both his undoing and his making. Humbled, blind, old, and outcast, Oedipus is a far nobler creature than he had been at the height of his kingly power. Tragedy, the Greeks thought, was necessary to temper our pride, to rein in the tug of hubris, and to open our eyes to hidden connections, obligations, and possibilities.

Kayford Mountain is surely a lot closer to New Orleans than the map shows. They are connected by the cycling of carbon in the biosphere that moves residues of ancient life to the outer atmosphere, where it traps the heat that amplifies storms in the Gulf of Mexico. They are connected by the lineaments of tragedy in which the poor are required to participate in their own undoing by ravaging their places in order to live. They are connected by the bonds of hubris through which some presume to tempt fate by violating the limits of the biosphere and thereby call forth suffering and misery. New Orleans and West Virginia are connected by the bonds of unnecessary suffering inflicted, it is said, by the necessity to have our energy cheaply and by the brutal imperatives of a kind of calculation that is used to justify endless wars while refusing to repair a wounded city and make its peoples whole.

We are now engaged in a global debate about what it means to become "sustainable." But no one knows how we might secure our increasingly tenuous presence on the Earth or what that will require of us. We have good reason to suspect, however, that the word "sustainable" must imply something deeper than merely the application of more technology and smarter economics. It is possible, and perhaps even likely, that more of the same "solutions" would only compound our tribulations. The effort to secure a decent human future, I think, must be built on the awareness of the connections that bind us to each other to all life and to all life to come. And, in time, that awareness will transform our politics, laws, economics, philosophies, manner of living, worldviews, and politics.[2]

CHAPTER 5

The Spirit of Connection

Only connect.

—E. M. Forster

Thinking means connecting things, and stops if they cannot be connected.

—G. K. Chesterton

Religio: to bind together.

—Webster's Dictionary

THE CONVERSATION ABOUT THE FUTURE OF HUMAN-kind and the preservation of life cannot be bottled up at the level of technology, economics, and politics, which have to do with means, not ends. In a vacuum of meaning and purpose, however, we don't do well either individually or collectively. Instead we are more likely to succumb to anomie, nihilism, and insensate violence. But questions about the purposes and the moral compass by which we might reorient ourselves have become much more complicated.

At the dawn of the 20th century, optimism about the human condition abounded. Science and technology seemed to promise an unlimited future, and in various ways larger questions were set aside in the intoxication with progress, the goal to master ever more of nature, and the hive-like effort to grow economies and eventually fight two world wars. But looking back across the wars, gulags, death camps, ethnic cleansings, killing fields, and mutual assured destruction, the 20th century appears rather like a passage through Hell. Looking ahead to rapid climate destabilization, the loss of perhaps a quarter to half of the species of life on Earth, and the widening gulf of poverty and living standards, we see that it may not have been a passage at all but a road toward the abyss of extinction. But it is a mistake, I think, to regard the possible suicide of humankind as an anomaly rather than the logical outcome of a wrong turn that now must be quickly undone.

For all of its complexity, the essence of the issue of sustainability was put by the writer of Deuteronomy long ago: "I have set before you life and death, blessing and cursing: therefore choose life, that thou and thy seed may live." No previous generation could make that choice as fully and finally as we can. We have the choice of life and death before us, but now on a planetary scale. One might expect that this choice would have been a matter of considerable interest to mainstream Christian denominations, but with a few notable exceptions they have been, in scientist Stuart Simon's word, "sluggish" to recognize such issues.[1] Were they members of a fire department, they would still be pulling on their boots as the ashes cooled. The same could certainly be said of other religions as well as other institutions, including those of higher education. But the problem of religion in America relative to the choices we face about the possibility of our own extinction is particularly important because of the close historical connection between Christianity and capitalism, which has been the engine of planetary destruction, and because of the rapid growth of an extreme branch of Christian fundamentalism that intends "to transform the church into the religious arm of conservative Republicans" and thereby "hijack faith and politics," in the words of liberal evangelist Jim Wallis (2005). The late Jerry Falwell, along with Pat Robertson, James Dobson, megachurch minister Rick Scarborough, Ralph Reed's Christian Coalition, the Southern Baptist Convention, and James Kennedy's Dominionists, have exerted great influence on U.S. politics and are said still to intend a fundamentalist takeover of the U.S. government "whatever the cost" (Moser, 2005). The result, in Bill McKibben's words, is that "America is simultaneously the most professedly Christian of the developed nations and the least Christian in its behavior" (2005). It is imperative, therefore, that we understand what extreme fundamentalists intend and what that portends for our democracy and our collective prospects.

Evangelicals and fundamentalists, however, are not all of one accord.[2] Fundamentalism, as historian of religion George Marsden points out is a "mosaic of divergent and sometimes contradictory traditions and tendencies" (2006, p. 43). The difference between evangelicals and fundamentalists, according to Marsden, is "their relative degrees of militancy in support of conservative doctrinal, ecclesiastical, and/or cultural issues" (p. 235). Many evangelicals, including Richard Cizik, the former vice president for government affairs at the National Association of Evangelicals, and Jim Ball of the Evangelical Environmental Network, are constructively engaged in "creation care," building alliances between churches and the environmental community to good effect. The problem, in the words of evangelical theologian Ronald Sider, is that:

> Tragically, Christian political activity today is a disaster. Christians embrace contradictory positions on almost every political issue. When they join the political fray, they often succumb to dishonesty and corruption... At the heart of the problem is the fact that many Christians, especially evangelical Christians, have not thought very carefully about how to do politics in a wise, biblically grounded way... [The result is] contradiction, confusion, ineffectiveness, even biblical unfaithfulness, in our political work. (2008, pp. 11, 19)

My concern, accordingly, is with the more-fundamental-than-thou brand of fundamentalism, and specifically with those at the extreme who profess belief in the imminence of the end times as allegedly foretold in the book of Revelation. In their view, there is no choice to be made between life and death because the Earth and all unbelievers are doomed anyway. The origins and evolution of this article of belief deserve closer scrutiny.[3]

A great deal of the purported Biblical justification underlying belief in the end times is due to the fertile theological imagination of a 19th-century Englishman, John Nelson Darby (Rossing, 2004, pp. 22–25). Darby's ideas, including the doctrine of dispensationalism, were later propagated in the *Scofield Reference Bible* distributed widely throughout the American South and subsequently broadcast far and wide by a herd of blow-dried electronic televangelists and propagated through the writings of Hal Lindsay, author of *The Late Great Planet Earth,* and the 12 volumes of Tim LaHaye's "Left Behind" series, which reportedly has sold more than 50 million copies (Rossing, 2004; Bawer, 1997). Darby's followers include the tireless merchants of fear and divine vengeance frothing on "Christian" radio across the heartland and theologians at right of right institutions such as Bob Jones University. After more than a century of considerable effort, the net result is that 83 percent of Americans say they believe the Bible to be either the literal or the inspired word of God (Harris, 2004, p. 230). Many of these also believe in the LaHaye doctrine of a final conflict between good and evil, the end times, the return of Christ, and the rapture that will gather the saved into the bleachers to watch the agony of the unbelievers burning down below in a lake of fire. There are helpful Web sites by which one can track precisely how close we are to the rapture, and bumper stickers warning that the car to which it is attached may be suddenly rendered driverless, creating difficulties for the passengers left behind and for insurance companies that will have to sort out liability issues.

There is also a more elite fundamentalism, hidden from public scrutiny, that has wormed its way by stealth into high places in government. In Jeff Sharlet's words, "elite fundamentalism, certain in its entitlement, responds in this world with a politics of noblesse oblige, the missionary impulse married to military and economic power. The result is empire. Not the old imperialism of Rome or the Ottomans or the British navy . . . Rather, the soft empire of America that across the span of the twentieth century recruited fundamentalism to is cause even as it seduced liberalism to its service" (2008, pp. 386–387). It is the fundamentalism of expensive prayer breakfasts in Washington, D.C., small cliques of the ardent and sophisticated in expensive suits constituting a net-work of "congressmen, generals, and foreign dictators who meet in confidential cells to pray and plan for a leadership led by God." The goal of what is known as "The Family" is the "maintenance of a social order through the salvation of souls," not the reform of the public order, or eradication of poverty, or improvement of the conduct of the public business (p. 382). Elite fundamentalists, like free marketers, focus on the individual, whether the rational actor of economic theory or the individual soul. But, as Sharlet points out, "Both deny possessing any ideology; both inevitably become vehicles for the kind of power that possesses and consumes the best intentions of true believers" (p. 383). And both have made their peace with the status quo and the powers that be.

A lot of religious nuttiness is loose in the land, and it might be dismissed as a matter of interest only to sociologists of religion and psychologists were it not for the fact that many practitioners have managed to become a large part of the political "base" of the Republican Party and of the most destructive U.S. presidency in history—one contemptuous of science, environmental pro-tection, international law, human rights, and world opinion, and much enamored of secrecy and the use of military power. During the Bush presidency, as David James Duncan puts it, "American fundamentalists . . . predominately support an administration that has worked to weaken the Clean Air and Clean Water Acts and gut the Endangered Species and Environmental Policy Acts; this administration has stopped fining air and water polluters, dropped all suits against coal-fired power, weakened limits on pollutants that destroy ozone, increased the amount of mercury in the air and water... the list goes on and on" (Duncan, 2005).

Both Ronald Reagan and George W. Bush, among other prominent Republican leaders, are said to have believed that these are in fact the end times, and even advanced the cause a bit. George W. Bush placed fundamentalists in positions of authority throughout the federal government, including departments and agencies administering federal lands and environmental laws, and these appointees were not shy about amending scientific reports in ways more agreeable to administration doctrine. Many professional environmental scientists and highly competent career civil servants were fired or forced into early retirement, replaced by others with apocalyptic religious views and considerable hostility to laws and regulations aimed to protect the environment and supportive of efforts to eliminate inconvenient regulatory barriers to resource extraction, pollution, and the preservation of species (Kennedy, 2004). And if, as Reagan's Secretary of the Interior James Watt is purported to have said, the end times are indeed upon us, there is a case to be made to exploit what's left of our forests, soils, and resources and close things out with a party. On the other hand, if the end is near, one might ask why bother to add another few percent to the gross national product? The faithful might better spend their remaining days toning up spiritually by prayer and fasting rather than laying up for themselves more treasure on a doomed Earth, which would be rather like winning another pot in a poker game on the *Titanic*. But extreme fundamentalism of any sort is not necessarily about consistency or even facts, but about maintaining the hold of doctrine on the faithful.[4]

The George W. Bush administration is now departed, and all of this would be so much ancient history except that its legacy remains and extreme fundamentalists are still with us, and by all signs still intend to change the nation into a theocracy. And even if somewhat subdued and hopefully in decline, the unholy alliance still exists between extreme fundamentalists and the vendors of fossil fuels, the climate changers, the polluters, the weapons merchants, the neocon imperialists, the exploiters, the political dirty tricksters who always believe that the ends they've chosen justify whatever means they use, the spin artists, those willing to corrupt scientific truth for political gain, and those for whom the law and the Constitution are merely scraps of paper. In the language of the book of Revelation, these are "the Powers" (Wink, 1984, 1986, 1992). Perhaps the alliance will founder on the fact that the members of this coalition have very different interests; extreme fundamentalists intend to bring about their version of the Kingdom of God on Earth, while their allies wish merely to establish American global dominance. Others simply want to make a great deal of money. As their part of the bargain, the Powers have been willing to say whatever extreme fundamentalists wish to hear about abortion, prayer in school, gay marriage, and flag burning, and will even appoint judges of their liking. They will attend prayer breakfasts and give stirring speeches professing their love of God while they loot the country by shifting taxes onto the middle classes and poor, move jobs overseas, undercut the laws that protect the environment and human health, wage wars in distant places using the sons and daughters of the poor as cannon fodder, and defame, threaten, or marginalize anyone who gets in their way, talking all the while about family values and morality. They have asked only that their supporters be blind, gullible, and ill informed, which is to say, ripe for the plucking.

Perhaps the influence of extreme religious fundamentalists on American politics is in decline, but I wouldn't bet on it just yet. Hard times tend to amplify, not dampen, extremism of all kinds, and our national brand of right-wing extremism is heavily infused with religious fundamentalism. Even so, some believe that saying such things is counterproductive because it will offend. Accordingly, politeness, not candor, should rule. But even within the context of Christianity, what particular style of discourse do they propose? Would it be that of Moses, who shattered the Ten Commandments at the feet of unrepentant Israelites? Or that of the Old Testament prophets, who called wayward people to task with unsparing honesty? Or that of Jesus, who tossed the money changers out of the temple with no particularly "constructive dialogue"? Or would it be that of Martin Luther nailing his 95 theses to the door of the Castle Church in Wittenberg? Or perhaps that of Dietrich Bonhoeffer, who railed against "cheap grace" and willingly died as a witness? Or is it the style of Martin Luther King, who spared no words to describe the connection between racism and the war in Vietnam? Or that of theologian and lawyer William Stringfellow, who once identified the United States as the heir to "the ethos and mentality of Nazism" (Stringfellow, 1973, p. 125)?

Said differently, at what point in the 1930s did the politeness of the German Church become the obsequiousness and then the full-blown cowardice that Pastor Martin Niemöller later famously lamented?[5] I do not know. But I do wish to inquire how best to converse with those much enthralled by the prospect of an end time, rapture, Armageddon, and the establishment of an American Taliban. Christians, evangelical and mainstream alike, have no business being in any alliance with the vendors of war, weapons, torture, corporate power, injustice, and ecological ruin (Wink, 1984, 1986, 1992). What is at stake now—the death of the ecological conditions that permitted humankind to flourish—calls for a higher level of honesty, directness, and spiritual wisdom sufficient to shift the perceptions, loyalties, and behavior of an entire nation. We should dialogue constructively when possible, but we must speak truth as clearly as we can see it and as unequivocally as we can say it.

If honest dialogue is not possible, should we perhaps dispense with religion altogether, as proposed by Richard Dawkins (2006) and others? Sam Harris, author of *The End of Faith* (2004), for one, proposes that "The problem that religious moderation poses for all of us is that it does not permit anything very critical to be said about religious literalism...[It] closes the door to more sophisticated approaches to spirituality, ethics, and the building of strong communities" (pp. 20–21). He intends to help "close the door to a certain style of irrationality...still sheltered from criticism in every corner of our culture" (p. 223). In his view, religious faith, unmoored from fact, data, logic, and the procedures of verifiability, poses a mortal danger to civilization. His book is rather like a stern reprimand for the foolish and dangerous religious thinking that has pervaded human cultures and now, with dispersion of weapons and means of mass destruction, threatens to undo civilization entirely. This is not the time, Harris writes, to preach tolerance of views that are patently disgusting, violent, and dangerous on a global scale, but rather a time to call religious extremists, Muslim, Christian, and Hindu alike, to account. Scientist Stuart Kauffman similarly proposes a nontheological sense of the sacred as "our own choice...in an emergent universe exhibiting ceaseless creativity" (Kauffman, 2008, p. 257).

I am not prepared, however, to toss the proverbial baby out with the bathwater. I believe that honest dialogue across our religious views is, in fact, happening, albeit cautiously and slowly. And there are points of obvious convergence, for example, between evangelicals and environmental scientists. One such is the observation that the doctrine of the end times bears a family resemblance to the views of some environmentalists—both agree that things are quickly going downhill. Earth systems scientists report almost daily scientific research documenting the unraveling of one system or another, habitat destruction, climate change, the ongoing loss of species, the effects of pollution, environmental threats to health, and the interaction of such factors as a larger cascade of bad news. No broadly informed scientist can be very sanguine about the long-term future of humankind without assuming that we will soon recalibrate human numbers, wants, needs, and actions with the requisites of ecology within the limits of a finite biosphere. Although one can quibble about the details and the schedule, most scientists are aware that climate change, biotic impoverishment, catastrophic pollution, resource wars, or a combination of all four pose a mortal threat to humankind. Evangelicals might regard the same evidence as a sign of the fulfillment of prophecy about the end times and the imminent return of Christ. But the differences between environmentalists and even some end-times fundamentalists are less than one might think (see table 5.1).

TABLE 5.1 Differences between environmentalists and end-times fundamentalists

	Environmentalists	Fundamentalists
Dynamics/drivers	Population Economic growth	Sin/evil/Satan
Source of problem	Lack of knowledge Policy failures Market failures	Human evil
Model	Ecosystems	Battleground: good v. evil
	Biosphere	Earth as fallen, heaven as true reality
Remedy	Better science Better policy Prices that tell the truth	Individual salvation Redemption Christ's coming
Tools	Government policy Economic policy Education	Christian education Redemption
Outcomes	Sixth extinction spasm, etc. Heat death of Earth	Armageddon, Second Coming...
Goals	Sustainability	Bring on the end times
Ironies	Little connection between deeper levels of human motivation and ecological problems	The end times become a self-fulfilling prophecy

Can one be an evangelical, for example, and a good environmentalist? Having known many who are both, I can say that the answer is "yes." But reconciling religious doctrine at the extreme with the goals of conservation requires some heroic intellectual acrobatics. On one side, belief that the end times are near tends to make extreme fundamentalists careless stewards of our forests, soils, wildlife, air, water, seas, and climate. It is a great deal easier to be concerned about conserving the Creation if one assumes that: (a) Earth is God's handiwork; (b) we are called to be good stewards of it and pass it on undiminished; and (c) humankind will be around for a while to enjoy nature and perhaps even be uplifted by it. Fundamentalists' belief in the end times, in other words, has the paradoxical effect of justifying behavior that brings on the end times, but of a sort with no authentic Biblical basis. The destruction of the Creation because of hard-heartedness and indifference to life is a sin against God and a crime against humanity. Further, careless talk about the imminence of Armageddon suggests a darker fascination with death, militarization, and violence discordant with the ideas of loving one's neighbor and the blessedness of the peacemakers.

Environmentalists, too, are in a quandary. For decades they have documented in great detail the decline of one life form or another and the destruction of ecosystems, but they are often tongue-tied on the deeper questions about the causes and forces driving the destruction of nature and climate change. They talk, instead, about changing economic policy, enforcing laws and regulations, adopting better technology, and improving science education, assuming that no further improvement of mind and spirit is necessary. Economist Herman Daly once said, "Afflicted with an infinite itch, modern man is scratching in the wrong place" (*Valuing the Earth*, 1996, p. 155). Until we do, however, the deeper motivations and the deeper causes of our ecological problems will elude us, and so, too, will the broad public support necessary to address them. Without a clearer understanding of why people seek the consolations of religion, we can make no persuasive case for sustainability beyond the fact that we wish to survive. In other words, environmentalists lack both a deep explanation for what ails us and a larger cosmology or spirituality rendered into a coherent and plausible alternative story of our ecological maladjustments. Neuroscientists and biologists are learning a great deal about what makes us tick, but can they tell a compelling, authentic, and life-oriented story of our human sojourn plausible and powerful enough to replace stories about the end times and the inevitability of Armageddon? If they cannot, charlatans will fill the human need for meaning in hard times with snake oil of one kind or another.

I do not presume to know what the story might be, but I think it must begin by placing science in a larger context where fact and mystery meet. There are many good examples of scientists who have been true to both science and its larger context, such as E. O. Wilson on biological diversity, David Ehrenfeld on humans in nature, Carl Safina on the oceans, Rachel Carson on the sense of wonder, Thomas Berry and Brian Swimme on "the universe story," and Stuart Kauffman on reinventing the sacred.

But within the scientific drama of life, evolution, and cosmos, who and what are we, and why do we, of all creatures, deserve to be sustained? Most of the debate about sustainability begins with the unstated assumption that since we want to survive, we ought to survive, making the question moot. But we should not let ourselves off the hook so easily. The reason I offer is entirely practical: if we could know *why* we ought to be sustained, we might better understand *how* to go about it. To know ourselves worthy of survival, for one thing, would lend energy to our efforts toward sustainability. If we believe ourselves unworthy—no more than maximizing creatures *Homo economicus,* or just clever primates—our efforts toward sustainability will lack the conviction that arises from knowing our cause to be valid. And knowing what makes us worthy of longevity will help us set priorities in the years ahead and determine those aspects of personhood, society, economy, and culture that ought to be preserved and those that must be jettisoned.

As a thought experiment along these lines, I once asked the students in my introductory environmental studies class to assume that they were to represent *Homo sapiens* before a congress of all beings, as once described in an ancient Islamic tale and more recently by Joanna Macy and Jonathan Seed (1988).[6] The charge against humankind would read something like this: "Over many thousands of years humans have proved themselves incapable of living as citizens and members of the community of life, and in recent centuries have become so numerous and so hazardous to other members of the community and the biosphere that they should be banished from the Earth forever." All the critters—reptiles, fish, birds, mammals, insects, and small things that make everything else work—are represented in the jury box and equipped with sentience and voice. The presiding judge is an owl, said to be the wisest of all; the prosecuting attorney is a fox, said to be the most cunning. The question for my students is simply: What defense might be made on our behalf? What supporting evidence could be presented? Who among the animals and plants would speak for us?

For the most part, students, while finding this an interesting exercise, conclude that no good defense can be made on any terms. But mostly, they stumble through the unreality of the scenario burdened by the assumption that humans are the pinnacle of evolution and that our desire to survive is a sufficient justification. Almost to a person they believe that, given our intelligence and the power of our technology, we will survive. A few believe that we are made more or less in God's image, giving us license to do whatever it is possible for us to do, devil take the hindmost. Otherwise articulate and intelligent, my students' confusion is representative, I think, of a larger befuddlement.

The case to be made against us is straightforward: we stand accused of being destructive, capricious, violent, wantonly cruel, derelict stewards, and unworthy of the appellation *Homo sapiens.* We are driving other species into oblivion and the Earth into a period of great and tragic instability. In his opening statement the fox states that: "Humans live beyond the limits and laws of nature and believe this to be their right. For every St. Francis, there are tens of thousands—no, hundreds of thousands—who are destroyers and killers, believing themselves exempt from the laws of community, decency, and courtesy, and millions more who give no thought to such things whatsoever. In fact, they are no more than rapacious and clever monkeys, but without the monkey's good judgment." Laughter erupts in the jury box; when it subsides the fox goes on. "Without the restraints of a small population, an all-embracing religion, law informed by nature, an ecologically grounded philosophy, technological incompetence, or even fore-sight, said to be their chief glory, humans are doomed, as some of their own have said, and deserve the death sentence before they take most of us with them. We must banish them from the Earth forever, and the sooner the better. I ask you, in the name of your children and your children's children, to sentence humankind to death at dawn." Members of the jury, excepting the cockroaches, mosquitoes, bedbugs, and kudzu, seem to mumble their assent as if in unison.

The defense has to contend with numerous complexities. Perhaps humans, for all of their protestations to the contrary, are haunted by a collective death wish, as Freud once thought. Perhaps we really are not so much a rational species as we are exceedingly clever rationalizers. Again, the evidence cannot be lightly dismissed. The sources of irrationality are many, starting with the still small voice of our genes that moves us to do their bidding and extending through our ineptitude at seeing patterns and systems and acting accordingly. Perhaps our evolutionary career has hardwired us to myopic tribal loyalties. Maybe we are just sinful and fallen, deserving of death excepting the redeemed, as fundamentalists would have it. Having multiplied extravagantly and extended our dominion over the air, seas, and lands and into the depths of the atom and gene beyond any rational limit, we are too successful for our own good. We define ourselves as consumers, a word originally designating disease. But what we consume is the planet's primary productivity, on which other species also depend. We think of ourselves as little more than rational players in an economic system conceived along with the industrial revolution 250 years ago—an infinitesimal slice of the 3.8 billion years of evolving life. The bloody catalog of history shows us to be stone-cold vicious against our own, against animals, and natural systems. The challenge to the defense is very large.

As the trial opens, the attorney representing humankind—for all of its cultural and scientific attainment, for all of its art, poetry, literature, and, yes, for all of its bloody history as well—rises to give her opening statement. Jurors in their various garbs of fur, fin, shell, and feathers lean forward to hear the defense.

"Most honorable Judge; my esteemed colleague of the bar, Mr. Fox; members of the jury: I am grateful for the opportunity you afford me to speak on behalf of my own kind, now facing charges that carry the gravest of penalties. I do so with fear and trembling for what the charges portend, but with confidence born of the knowledge that our species, for all of its shortcomings, is a worthy and promising part of the community of life. I appreciate the opportunity to speak to you members of the jury as a family with a long history. From the earliest stirrings of life in the seas, our path has been a long intertwining of biological destinies, of sharing genetic material, and of mutual learning. We have even been food for many of you." Jurors, except the parasites and those with fang and claw, looked baffled, unsure whether this was an ill-conceived attempt at humor or something darker. The shark shows no emotion at all.

The attorney for the humans continues. "We have learned much from each of you. Our first inkling of what we are was shaped by communion with you, Mr. Bear, and you, Mr. Wolf, and you, Ms. Salmon—indeed with all of you. We first came to know many of you as our teachers—the mirror by which we might better understand ourselves. For reasons not our fault, we are the only species troubled by self-consciousness and the knowledge of our mortality ... a burden that weighs far less on all of you. Our first art attempted not just to portray some of you, but to honor you for what you taught us about ourselves. Many of you graciously fed us when we were hungry. Many of you fed our spirits by your ability to soar in the skies or play in the waters. You taught us faithfulness to place and seasons. You taught us industry, thrift, and the determination necessary to survive. The trickster coyote taught us cunning when we were weak. Our first words were a crude imitation of the sounds some of you make. You taught us the habits of work and even the arts of making nests, dams, and homes. You taught us to fly, to swim, to navigate, and to return home again. You were our first teachers, for which we are grateful. Had we been more adept students we would have better learned the arts of managing fertility and sunlight taught by our sisters, the forests, the grasslands, and the deserts. Nonetheless, what we are now owes much to those early lessons mastered all too imperfectly. But we are quickly learning how to better mimic your ways in our own industries." The jurors stir ominously.

Undaunted, the attorney for humankind proceeds to her next point. "We have evolved together on this small beautiful planet. But neither we nor most of you are what you once were thousands or millions of years ago. Excepting a very few of you, such as you, Mr. Horseshoe Crab, we have all changed. Even so, we show the unmistakable signs of our common origins in the seas. Humans differ only slightly in the makeup of their genes from our kin of only a few tens of thousands of years ago, a mere snap of the fingers in time. Still there is a difference, our mark, as it were. Each of you jurors has a specialty shown by fang, appendage, or power of sight or speed or disguise. Humans are generalists endowed with minds capable of language, reason, abstraction, and sufficient

foresight to fear our own demise. None of you can do what we
can do, and none of you carries the fearful knowledge of mortality
that we bear. But that knowledge came with an obligation as well,
for it was left to us to give voice to the journey of life on Earth; to
write its poetry, paint its pictures, fathom its meaning, and ponder
its ascent and final end—to ask why and how. Knowledge, we
now understand, is both liberating and damning. Why this was
left to us, and to none of you, no one can say. And no one can
say what knowledge will do to humankind as the millennia roll
forward. All of us in this courtroom are in a slow transition from
what we were and what we are to some unknown future. The
particular advantage of my kind is the mental capacity to learn
and create culture much faster than the evolution that shaped all
of you in this courtroom. The transition of which I speak is gathering force and speed."

The jurors are restless, impatient with what appears to be an
irrelevant diversion. The wolf can be overheard muttering to the
elephant that "as they steal more of our secrets, the enemy," as
he puts it, "will become even more tyrannical and destructive."
The elephant makes no response. Mr. Fox rises to address the
judge. "Your honor, this line of argument is immaterial to the
charges at hand. I respectfully request that the defending attorney
be instructed to get to the point, and quickly." He sits. The jurors,
growing impatient, nod in agreement. The defense attorney stands
and responds: "Your honor, I respectfully submit that this is most
relevant, and I will shortly explain why and how." In a flat voice
the judge snaps: "Proceed, but be quick about it."

"Thank you, your honor. To you members of the jury I will
offer no justification for past wrongs, excesses, and cruelties
inflicted on you and your ancestors by my own kind. But I do ask
each of you to carefully consider the evidence that I will present
of what is happening all around you. All over the Earth a great
turning in the evolution of humankind has begun. It is driven
by the forces of which I spoke moments ago. Our capacities to

learn, reason, and even empathize are growing quickly. We now
know ourselves to be a part of a larger story of life in the universe
and are beginning to understand what that will require of us. All
over the Earth humans are engaged in a momentous conversation
about the terms and conditions that must be met in order to sustain life—yours and ours—on this planet.

"A word about our own history is in order. Cruelty toward
our kind was part of that history. After many years, however, and
with much trouble, we have learned the value of law, restraint,
fairness, decency, democracy, and even peace. Not long ago, one of
my gender could not have been selected for the heavy responsibility that I now bear. Have we learned these lessons well enough?
By no means! But they now represent a growing force in human
affairs, spread by our global communications technology. We now
know instantly of problems and crises that occur all over the
planet, including news of our own folly. Do we always respond
adequately? By no means! But we are learning, and most important, millions of people now consider their allegiance to Earth, to
the future, and even to all of you as members of the community
of life to be more important than those to nation and religion.
Is the battle for decency won? No. But in time, I submit that it
will be."

Members of the jury, if not mollified, appear to be less hos-
tile. But the wolf, leaning on the rail of the jury box, shows utter
contempt.

The attorney for humankind continues. "As I will show, humans
are the first species to show kindness to another species. We, not
you, ponder and often worry about such things as justice, fairness,
and decency, not simply the laws of eat and be eaten. Nothing in
nature dictates such things, but we believe this, too, a part of our
obligation to the community of life. We have laws, imperfect to
be sure, protecting each of you in some fashion." The rats, mice,
chimpanzees, and a few other subjects of laboratory experiments
exchange angry glances. "We are the first to see Earth from space,

measure its temperature, count the number of species, and understand its laws. We are the first among all of Earth's diverse life
forms to understand our world enough to take steps to protect
it." A member of the salmon nation shouts in response that "it
would not need protecting were it not for your kind!" Shouts
erupt throughout the court. The judge calls for order.

The attorney for the defense resumes. "The angels of our better nature are growing more powerful in human affairs. There is
now a global movement to protect species, stabilize the climate,
preserve habitats for each of you, rein in our excesses, and reduce
consumption. Efforts have begun to restore lands and waters that
we have degraded through carelessness and ignorance. We are
learning the arts of designing with natural systems in ways that
give back as much as they take. We are beginning the great transition from coal and oil to efficiency and sunlight. If granted the
right to survive, the difficulties and challenges we face in the years
ahead will be many, but the great turning in human attitudes and
behavior has begun. We, a young species compared to many of
you, are beginning to fulfill our promise for wisdom, compassion, and foresight. We are acquiring the scientific and technological know-how necessary to radically reduce our impacts on the
Earth."

"For all of our shortcomings and liabilities, I ask you to ponder
not just a world without humans...." She is cut off by shouts of
"We'd like to!" and laughter. She resumes, slowly measuring each
word: "...but a world not far into the future of a partnership of
life on Earth, of mutual celebration between evolution and intelligence—a better world for all. I ask each member of the jury to
see this as dawn, not sunset; a beginning, not an end." Her opening statement finished, she sits. The judge asks the attorney for the
prosecution to call his first witness. The trial begins.

I asked my students to consider how the trial might turn out,
and why. Is there something special about *Homo sapiens* that trump
other considerations? Is there a better defense than one based on a

promise to improve? Is there any evidence that we are doing better or that we will do better? Is there a kind of middle sentence between life and death? Under what terms could humankind receive a contingent life sentence or probation?

The trial, like philosopher John Rawls' "veil of ignorance," is a heuristic device to help us see what we might otherwise miss. But it is more than that. It is an invitation to ask those age-old questions, now more important than ever, about what we are and where we are going.

There will, of course, be no trial, no parole, and no contingent sentence, only an eerie and deepening silence as species disappear—unless and until we shift course. As my students know as well, there are profoundly important efforts under way to change our course along with formidable sources of resistance and the brute inertial momentum of industrial civilization. The difference between these outcomes depends to a great extent on whether humanity is capable of quickly learning new behaviors appropriate to a planet with a biosphere and a higher vision of what we could become. Is there in us a promise of something more? Perhaps we have, as Joel Primack and Nancy Abrams suggest, a "sacred opportunity...a chance to be heroes...to become the kind of people capable of using science to uphold a globally inclusive, long-lived civilization" (2006, pp. 295–297). But science on its own won't save us in the absence of a renewed sense of the sacred sufficiently powerful to overcome our indifference to Earth, which is to say absent a change of heart.

Perhaps we are not alone in this effort. Philosopher and theologian Thomas Berry, for one, believes that "We are not left simply to our own rational contrivances [but] are supported by the ultimate powers of the universe" (1988, p. 211). In contrast to the story of human conquest and progress, Berry and cosmologist Brian Swimme propose that we better fit a different and larger narrative in which humankind is a part of a still-evolving universe (Swimme and Berry, 1992). Against the vastness of cosmic evolution, the "universe story" diminishes our pretense of mastery while raising the importance of humankind as the storyteller in an otherwise silent universe. Theological details aside, I believe that we need to see ourselves as part of a larger story. But what story?

Several years ago I was asked to deliver a college baccalaureate address, which is not my custom, and foolishly agreed. In trying to get traction on the subject I reread Berry and Swimme and others and came to the conclusion that we need a larger story for certain, but we also need to begin with something closer to hand and heart, which is simply the sense of gratitude for the gift of life itself.[7] The address, my first and last attempt at theology, went like this.

The story of the universe begins in the great cymbal smash of the Big Bang, the rhythm heard through the still expanding Creation, and in the pulsations of energy and light that animate the cosmos and drive the journeys of our little planet around its small star. Day follows night; one season follows another in the dance of life, the ebb and flow of the tides, the migration of birds, the rhythms in our bodies, and the seasons of our lives. That rhythm was reflected in early scripture and mythology. There is:

> A time to be born, and a time to die; a time to plant, and a time
> to pluck up that which is planted; a time to kill and a time to heal;
> a time to break down and a time to build up; a time to weep, and
> a time to laugh; a time to mourn, and a time to dance; a time to
> cast away stones, and a time to gather stones together; a time to
> embrace and a time to refrain from embracing.

It was once known that if we broke the rhythm, our little part of the cosmic dance would stumble to a halt. A fraction of a second ago, as geologists and ecologists measure time, another rhythm was begun. Some call this the Fall. In one telling of the story, the cadence was changed by a snake and a woman—a libel against a perfectly fine life form and against all women. More likely the discord came from a few males who thought that an elite few could improve the creation by changing the rhythm. C. S. Lewis once said that the intent was to control other men by seizing control of nature. Ecologically, control meant exploiting the vast pools of carbon—first the carbon-rich soils of the Fertile Crescent, later the carbon in the forests of Europe, and in our time the ancient carbon stored as coal and oil.

But it was not long before others, more sophisticated and clever, realized that a few could change the rhythm of Creation altogether. The heroes of disharmony, men like Bacon, Descartes, and Galileo, taught us that we could and should conduct the symphony. And so instructed, we learned how to make things never made by nature, we learned to split the atom and to manipulate the code of life. Some are busy making devices that will be, they say, more intelligent than humans. In the conquest of nature and of other men, the rhythms changed to those of the business cycle, the product cycle, the electoral cycle, the seasons of fashion and style—the rhythms of commerce, greed, power, and violence. But we did not know what we were doing, as Wendell Berry once said of the European conquest of America, because we did not know what we were undoing.

Now we live in a time of consequences. Climate scientists have given us an authoritative glimpse of a literal hell not far into the future that will change the seasons, the cycles of nature, the rhythms of life, and the great procession of evolution. We now march to the cadence of hubris, greed, and violence—and we should ask why.

Was it our capacity for denial, as psychologist Ernest Becker once said? Or our will to power, as Nietzsche thought? Or some flaw in our mental facilities? (There are some who half-jokingly believe that the next time God runs the experiment, the frontal lobe of the primate brain should be left out.) Or as an unwary species, did we simply trap ourselves in what Ronald Wright calls a "progress trap"?[8] I have come to believe, however, that the great Jewish rabbi Abraham Heschel had it right that the source of dissonance is ingratitude. As civilization advances, Heschel wrote:

> the sense of wonder almost necessarily declines…mankind will not perish for want of information; but only for want of appreciation. The beginning of our happiness lies in the understanding that life without wonder is not worth living. What we lack is not a will to believe but a will to wonder." (Heschel, 1990, p. 37)

The problem as Heschel puts it is simply that "A mercenary of our will to power, the mind is trained to assail in order to plunder rather than to commune in order to love" (p. 38).

We were given the keys to paradise but presumed that we could improve it on our terms, and so reduce the great mystery of life to a series of solvable problems, each contained in one academic box or another. We thought that we could rid ourselves of reverence and so exorcise mystery, irony, and paradox. We thought that we might change the cadence of Creation and seize control of the great symphony of life with no adverse consequence. We lost the capacity for gratitude.

But why is gratitude so hard for us? This is not a new problem. Luke (17:12–19) tells us that Jesus healed the ten lepers, but only one returned to say thank you. That's about average, I suppose. In colleges and universities, we teach a thousand ways to criticize, analyze, dissect, and deconstruct, but we offer very little guidance on the cultivation of gratitude.

And maybe we should not be grateful. In the spirit of pluralism, is there a case for ingratitude? Is gratitude merely a ploy that runs inversely proportional to favors not yet granted? One might suspect the Psalmist of such. Or perhaps there is no cause for gratitude amidst the cares and trials of life. Shakespeare, for example, has Macbeth say that "Life's but a . . . tale told by an idiot, full of sound and fury, signifying nothing." Political philosopher Thomas Hobbes similarly thought that life was full of peril and death: "nasty, brutish, and short." Shakespeare and Hobbes are Englishmen both, and so it is possible that gloomy weather had something to do with their opinions. But many of us find our bodies, incomes, careers, and lives to be less than we would like, whatever we might really deserve. Nevertheless, most of us, too, would find living life without appreciation rather like eating a meal without flavor or living in a world without color, or one without music.

So we set aside one day of the year for Thanksgiving, but mostly spend it eating too much and watching football. Gratitude comes hard for us for many reasons. For one thing, we spend nearly half a trillion dollars on advertising to cultivate ingratitude, arguably the source of the seven deadly sins.[9] The result is a national cult of entitlement, the desire to have as much as possible for doing as little as possible. For another thing, the pace of modern life leaves little time to be grateful or awed by much of anything.

But there are deeper reasons for ingratitude. Gratitude does not begin in the intellect but rather in the heart. "Intellect," in Brother David Steindl-Rast's words: "only gets us so far... It is the task of the intellect to recognize something as a gift" (1984, p. 13). To acknowledge a gift as a gift, however, requires an act of will and heart; to acknowledge a gift is also to admit "dependence on the giver... but there is something within us that bristles at the idea of dependence. We want to get along by ourselves" (p. 15). To acknowledge a gift is to acknowledge an obligation to the giver. And herein is the irony of gratitude. The illusion of independence is a kind of servitude, while gratitude—the acknowledgement of interdependence—sets us free. Only "gratefulness has power to dissolve the ties of our alienation," as Steindl-Rast puts it. But "the circle of gratefulness is incomplete until the giver of the gift becomes the receiver: a receiver of thanks... The greatest gift one can give is thanksgiving" (p. 17). To say "thank you" is to say that we belong together—the giver and thanksgiver—and it is this bond that frees us from alienation.

And the gift must move. What is given must be passed on. In the end, nothing can be held or possessed—a truth grasped by every culture that approaches what we've come to call sustainability. And in reciprocity, gratitude changes the rhythm. It restores the cycle of giver and receiver and back again. It extends our awareness back in time to acknowledge ancient obligations and forward in time to the far horizon of the future and to lives that we are obliged to honor and protect. Gratitude requires mindfulness, not just intelligence. It requires a perspective beyond self. Gratitude is at once an art and a science, and both require practice.

The arts and sciences of gratitude, which is to say, applied love, are again flourishing in ironic and interesting ways in places least expected. Businessman Ray Anderson has set his company on a path to give back what it takes from nature, operating by current sunlight and returning no waste to the Earth. Biologists are developing the science of biomimicry, which gratefully uses nature's operating instructions evolved over 3.8 billion years to make materials at ambient temperatures without fossil fuels and toxic chemicals, rather like spiders that make webs from strands five times stronger than steel. The movement to power civilization from the gift of sunshine and wind is growing at 40 percent per year worldwide. The American Institute of Architects and the U.S. Green Building Council have changed the standard for buildings to eliminate use of fossil fuels by 2030. These cases and others illustrate something that goes beyond mere practicality. The persons involved, I think, are animated by a deep sense of appreciation for the beauty of the world and the desire to pass it on intact to those who will follow us.

Could we, in time, create a civilization that honors the great gift and mystery of life? The "Great Work" of our generation is to stabilize and then reduce greenhouse gases, build a world powered by efficiency and sunlight, stop the hemorrhaging of life, and work for a time when every child is well loved and well cared for. Like previous generations in times of peril, none of us asked for these challenges. But it has been given to us to lay the foundation for a durable and just global civilization, to secure the gift of life and pass it on undiminished to unnumbered generations. No previous generation could have said that, and none ever had greater work to do.

Can true gratitude transform our prospects? Can we harmonize the rhythms of this frail little craft of civilization with the pulse of the universe? I believe so, but gratitude cannot be legislated or forced. It will remain a stranger to any mind that lacks compassion. It must be demonstrated, but above all it must be practiced daily. Our generation and no other has been given this Great Work, and for that we can be grateful and humbled. In that work, our strongest ally will be grateful hearts.

On closing, I asked each person in the audience to say "thank you" to someone present to whom they owed an unacknowledged debt. What I thought would be a few minutes turned into 20 minutes or so of fervent greeting, hugging, and even a few tears. The experience led me to think that we, in this affluent society, live mostly in a poverty of gratitude. And that is a simple problem to solve.

PART III

FARTHER
HORIZONS

CHAPTER 6

Millennial Hope

The gods and demons have not disappeared at all, they have merely got new names.

—Carl Jung

There is but one way to save ourselves from this hell: to leave the prison of our egocentricity, to reach out and to one ourselves with the world.

—Erich Fromm

I am unwilling to believe that this whole civilization is no more than a blind alley of history and a fatal error of the human spirit.

—Vaclav Havel

FOR TWO CENTURIES AND LONGER, HUMANKIND HAS been on a collision course with the limits of the Earth. The inertial momentum—the scale and velocity of the human enterprise—has grown so rapidly since the mid-20th century that virtually every indicator of planetary health is in decline (McNeill, 2000). Even an otherwise self-characterized "optimistic" analysis concludes that:

> The momentum toward an unsustainable future can be reversed, but only with great difficulty. [The reversal] assumes fundamental shifts in desired lifestyles, values and technology. Yet, even under these assumptions, it takes many decades to realign human activity with a healthy environment, make poverty obsolete, and

ameliorate the deep fissures that divide people. Some climate change is irrevocable, water stress will persist in many places, extinct species will not return, and lives will be lost to deprivation. (Raskin et al., 2002, pp. 94–95)

Considerably less optimistic, Thomas Berry concludes that "It is already determined that our children and grandchildren will live amid the ruined infrastructures of the industrial world and amid the ruins of the natural world itself" (2006, p. 95). James Lovelock's view is even darker: "the acceleration of the climate change now under way will sweep away the comfortable environment to which we are adapted.... [There is evidence of] an imminent shift in our climate towards one that could easily be described as Hell" (2006, pp. 7, 147; *The Vanishing Face of Gaia,* 2009). Given such dire predictions, theologian Jack Miles, author of *A History of God* (2000), suggests that we begin to ponder the possibility that "the effort to produce a sustainable society has definitively failed...that we are irreversibly en route to extinction." Alan Weisman, in a striking exercise of journalistic imagination, describes in *The World Without Us* how our infrastructure would then crumble, collapse, and finally disappear (2007). These are only a few of the recent musings about the human prospect. But we've been alerted, warned, and warned again by ecologists, geologists, systems analysts, physicists, sociologists, political scientists, biologists, National Book Award winners, Pulitzer Prize winners, Nobel laureates, teams of international scientists, and the wisest among us, but so far without much effect.

Might we still avert catastrophe? Facing the realities of ecological decline, the end of the era of cheap fossil fuels, and the destabilization of climate, it is not easy to find solid ground for hope in a sea of wishful thinking, evasion, and half measures. I believe that there are grounds for genuine hope, but since we have frittered away our margin of safety, they are a century or more ahead in an unknown future when we have stabilized the carbon cycle, reduced the concentration of greenhouse gases to preindustrial levels, stopped the hemorrhaging of life, and finally ended the curse of violence, and when the biosphere has begun to self-repair. On the other side of E. O. Wilson's "bottleneck" we do not know what size the human population will be, what history our descendants will have traversed, how much biological diversity will have survived, or whether stressed ecosystems will recover in time spans meaningful to humans (Wilson, 2002). James Lovelock, for one, believes that the human population will be no larger than one billion, and likely much less (2006, p. 141). One-third to perhaps one-half of the species presently on Earth could go extinct in this century. Heat stress, changes in the amount and intensity of rainfall, and ecological degradation will drastically change most ecosystems.

The Earth, then, will be very different from the planet we've known. Our descendants who come through the bottleneck may reside in the same places in which we do, but they will most likely live in very different circumstances than we presently do. They will be the survivors of a close call with extinction. Will they know, and, if so, how will they understand that history? What will they make of the ruins of industrial civilization, some submerged far beyond different shorelines? What kind of people will they be? Will they understand the events, trends, processes, and people who took us to the brink of extinction? What will they know about the pre-bottleneck world? Will they know what they were denied? Will they have succeeded in preserving the best of human culture, literature, and art? Will they live in a democracy, a totalitarian state, or tribal anarchy?[1]

The reality is simply that the planetary destabilization we've caused and all of the collateral damage to civilization will grow worse before the Earth's systems stabilize—hopefully with a biosphere still capable of supporting civilization. In the meantime, there will be contrary trends: wind power will continue to grow rapidly, new technologies will become available, businesses and other organizations will make great changes in the way they do business, the forces of blessed unrest will become still more restless, and governments and international agencies will finally bestir themselves to do belatedly what they should have done decades before. But these trends, as important and urgent as they are, will not quickly reverse the effects of climate destabilization to which we are now committed. Lacking foresight, we did too much damage to the fabric of life and waited too long to reverse the trends. For a while at least, our plight will be rather like that of the passenger walking north on a southbound train.

At the end of the era of cheap fossil fuels and climate stability, however, the forces of denial embedded in our politics, media, education, and economy will try to divert our attention as long as possible. Others will promise increasingly heroic ways to keep climate destabilization at bay by proposing one silver-bullet solution or another. For some, nuclear power is our only option, but they do not say whether it could be deployed on the scale necessary in the short time available and at a price we could afford. Nor do they explain how to manage terrorist threats or the radioactive waste that will have to be isolated for 250,000 years, or why this is preferable to improved efficiency and renewable energy technologies that can be deployed more quickly at a fraction of the cost and with virtually none of the risks and problems of nuclear energy. Others aim to develop and deploy devices that suck carbon out of the atmosphere (Broecker and Kunzig, 2008; Homer-Dixon and Keith, 2008) or cool the Earth temporarily by injecting sulfur dioxide into the stratosphere. If such geoengineering schemes are successful and if they do not cause other problems, they could buy us a bit of time, but little is said about how we might best use that short reprieve (Mooney, 2008). There will be other proposals, no doubt, and they will be increasingly grandiose and desperate; sometime soon one or more may be tried, and if so, they will likely fail grandiosely at an exorbitant cost (Robock, 2008, pp. 14–18). Despite their differences, these schemes all share the assumptions that the Earth is a machine and that it can be fixed by other machines, a variant of the kind of thinking that got us in the mess in the first place. As with earlier technological fixes, they will incur unanticipated consequences that will create still other problems to be fixed by still more machines, at a great profit to large organizations that will bear no part of the liabilities arising from consequences of their making. Geoengineering, salvation by gadgetry, rests on the belief that we are incapable of better behavior, learning, foresight, sacrifice, or exercising wisdom about the long-term consequences of what we do. In the words of one analyst, "No one knows today whether geoengineering could ever make sense" (Kunzig, 2008, p. 55). It is clear, however, that at best it is a temporary patch on a deeper problem whose solution begins by changing the way we harness and use energy.

Even in the most optimistic scenario, however, no known technology could be deployed on the scale necessary in the time available to avert the gathering storm. The conclusion is unavoidable: a great deal that we have long taken for granted, like the Sunday drive, the trip to the mall, the SUV, cheap food on well-stocked store shelves, or even the transition from one season to another, will likely become intermittent, or perhaps even distant memories. As climate destabilization bears down on us, there will be a lot less talk about economic growth, progress, individualism, empire, and defense of "the American way of life." These concepts will some day be no more useful to us than the ghost dances once were to the Plains Indians in halting the tide of white civilization sweeping across what was once their land.

Given the evidence is that the road ahead will be longer and more difficult than our leaders have been willing to admit, and likely longer and more difficult than they understand. But our immediate steps are clear: preserve soil and forests, save species, use less, deploy solar technologies, throw the rascals out, demand accountability in government and business, elect leaders with the courage and intelligence to lead in the right direction, and shift the center of American politics neither left nor right but from the status quo toward a livable and decent future. We must also contend with the defects in culture, politics, science, and society that caused the problem in the first place. The modern project—Promethean in its ambitions and Cartesian in its methods—has, on balance, turned out badly: a reality inadequately described by the vague and sterile word "unsustainable." For whatever short-term good it brought in the end, it is a kind of protection racket, in the poet Gary Snyder's words:

> economically dependent on a fantastic system of stimulation of greed which cannot be fulfilled, sexual desire which cannot be satiated and hatred which has no outlet except against oneself [or] the persons one is supposed to love. . . . All modern societies [are] vicious distorters of man's true potential. They create populations of "preta"—hungry ghosts, with giant appetites and throats no bigger than needles. The soil, the forests and all animal life are being consumed by these cancerous collectivities. (Snyder, 1969, p. 91)

Climate destabilization, in other words, is only a symptom of a much deeper problem.

A great deal now depends on what we do to develop the stamina, vision, and institutional resources necessary to carry the best of civilization through to the other side. Immediate action to reduce carbon emissions and preserve ecosystems will reduce the scale and duration of the traumas they will otherwise experience. Every effort to build local resilience and sustainable communities that meet many of their own needs for food, energy, water, and livelihood will minimize many of the risks people and all creatures will face. Changes in education to equip children and young adults with an understanding of the skills of self-reliance and ecological design will enable millions to participate directly in the making of the post-fossil-fuel world. Efforts to restore democracy, restrain corporate power, reinstate public control over common property assets, establish the rights of posterity to life, liberty, and property, and build a global community based on nonviolence, law, and fairness will create the political foundation necessary for recovery. These are the first steps of what Snyder calls a thousand-year journey. All of this is to say that genuine hope is neither passive nor resigned. To the contrary, hope means putting aside all of those traits of mind and character that prevent us from getting down to work with ingenuity, persistence, and good heart while understanding that the journey will be long and difficult. And there's the rub.

Are we, in the main, the kind of people who can face difficult realities and not flinch? Can we overcome the tendency to settle for half-truths and evade the reality settling in about us? In short, do we have the collective intelligence, courage, stamina, and heart to surmount the challenges ahead? No one can say for certain. What can be said is that our best chance to accomplish the great work ahead depends on a deep understanding of our potentials for good and evil and the cultivation of our higher capacities for wisdom, foresight, and altruism. Who are we and what do we know of ourselves? To successfully navigate the decades and centuries of the long emergency will require that we answer that question without illusions, but also without selling ourselves short.

Much of the talk about the challenge of sustainability skips around the nastier side of our nature. That evasion misleads us to think that we can get off the hook cheaply with only a little more cleverness. Against that view, others are beginning to see the impending crisis of climate destabilization as primarily a matter of morality, not economics or technology (Hillman, Fawcett, and Rajan, 2007, p. 243; Gelbspan, 2004, p. 181; Garvey, 2008). But that does not make our choices easier; to the contrary. As Nietzsche argued in *The Genealogy of Morals,* "morality will gradually *perish* now," and the history of the subsequent century seemed to confirm his pessimism as we stumbled from the slaughter of World War I to Auschwitz, Dresden, the Gulag, Hiroshima, Vietnam, the killing fields of Cambodia, Rwanda, and Darfur. After a lifetime of observing human nature, Carl Jung concluded that "There is a terrible demon in man that blindfolds him, that prepares awful destruction" (quoted in Jarrett, 1988, p. 1277). Historian and sociologist Barrington Moore in his classic study *Reflections on the Causes of Human Misery* concluded similarly that "mankind can expect to oscillate between the cruelties of law and order and the cruelties of changing it for as long as it leaves the globe fit for human habitation" (Moore, 1972, p. 39). In *Humanity: A Moral History of the Twentieth Century*, Jonathan Glover doubts that we became worse than previous societies, but says: "Technology has made a difference. The decisions of a few people can mean horror and death for hundreds of thousands, even millions, of other people" (p. 3). Looking beyond the carnage of the past century, he proposes to defend the Enlightenment hope of a more humane world, but concludes that "there are more things, darker things, to understand about ourselves than those who share this hope have generally allowed" (p. 7). He concludes by saying: "It is too late to stop the technology. It is to the psychology that we should now turn" (p. 414). Philosopher Tzvetan Todorov similarly believes that the nature of evil has not changed, but its scale has grown, driven by both fragmentation and depersonalization (Todorov, 1996, pp. 289–290).

Now that we are facing our largest challenge, what do we know of our own liabilities and potentials? It is easier to ignore that kind of introspection, to focus on technology or policy or anything else instead of looking inward to the complexities and ironies of our own psychology. But I think Glover is right that the job of building a decent world will come down to how well we understand ourselves and how much we can improve the "still unlovely human mind" (Leopold, 1949). The failure to do so explains in large measure why many underestimate the scope and scale of the human destruction of nature and trivialize its causes. As if caught in a bad dream, we seem powerless to stop it. The rapid destabilization of climate and the destruction of the web of life are symptoms, in part, of a prior derangement in our manner of thinking and in our ability to think clearly about how we think. So what is known about the mind that would be useful to enhancing our longer-term prospects?

The human mind is certainly capable of great feats of imagination and invention, as well as less agreeable behavior. It is both the crowning distinction of humankind and our greatest perplexity and liability. Mind reflecting on itself has long been a source of amusement, philosophy, and more recently, science. Now it is a matter of survival and of the terms and conditions by which we will survive through the long emergency.

For one thing, we know that people go to considerable lengths to maintain a favorable self-image and deny unpleasant truths, particularly those that run against deeply ingrained beliefs and worldviews (Allport, 1954).

We know that perception is biased toward the near term (Ornstein and Ehrlich, 1989). We tend, accordingly, to see things that are large and fast but not those that are small and slow. It is harder for us to see and to properly fear long-term trends, such as soil erosion over centuries or the nearly invisible disappearance of species. Similarly, in spite of centuries of modernization, our loyalties are most strongly attached to those closest to us. We bear the unmistakable signs of our distant origins as tribal peoples telling each other stories around ancient campfires.

We know, too, that we are prone to deny uncomfortable realities at both the personal level and the societal level. Like Voltaire's Dr. Pangloss, we tend to believe that things always work out for the best. I recently asked a class of U.S. college students, for example, how they would define climatic change. After some discussion they reached a typically American consensus that it should be defined as "an opportunity." They were not clear exactly how the opportunity would manifest at various increments of warming for exactly whom, and I did not ask how the opportunities would work out for those now in the crosshairs of rising seas, larger floods, bigger storms, prolonged droughts, and searing heat waves. These are smart students, but they reflect both the optimism of youth and the deep tendency to deny unpleasant things now amplified by a culture moving at warp speed. We, particularly in the West, are inclined to interpret all difficulties and impediments as merely problems that are by definition solvable with enough money, research, and technology.

For another thing, we know that people often hold two contrary beliefs at the same time and remain happily oblivious to the contradictions. Psychologists call this "cognitive dissonance" (Festinger, 1957). The trait manifests itself among those whose professed creed requires loving one's enemies while at the same time zealously bombing the hell out of them, without the slightest twitch of confusion. Having invested in an opinion or worldview or having made a particular decision, we go to considerable lengths to maintain the investment. "Confronted with dissonant information," as Tavris and Aronson put it, "the reasoning areas of the brain virtually shut down" (Tavris and Aronson, 2007, p. 19). Confronted with evidence of climate change, deniers exhibit classic symptoms of cognitive dissonance. But others, who admit the reality of climate change, often do the same by denying the severity of the crisis. Humans, perhaps, are not so much rational creatures as very proficient rationalizers.

Psychologists know, too, that we are prone to conform to peer pressure and group opinions even when those defy the evidence of our own senses (Asch, 1955). It is the same trait described in the story about the child who noted the emperor's nakedness while the adults denied the naked truth of the matter. The pressures to conform can work to preserve our nastier traits, such as racism or dubious opinions about climatic change, as long as they fit the group opinion. But they can also help to preserve our better traits, once they are integrated into the larger culture.

Conformity to group pressures in combination with uncritical acceptance of authority can be a particularly powerful determinant of behavior, causing people to do heinous things to each other. In a well-known study, psychologist Stanley Milgram put subjects in situations where actors posing as experts asked them to deliver progressively higher electrical shocks to supposed victims (1969). Under the authority of fake scientists, most agreed to do so despite the screams and simulated suffering of the actors posing as victims. In real life there is lots of evidence that ordinary people under duress can behave similarly. But sometimes they don't.

Elite decision making has its own pathologies. In his study of the Cuban missile crisis, Janis (1972) showed how the centripetal pressures of "groupthink" can deform decision making by narrowing perspectives and limiting the permissible evidence and imagination. In the Bay of Pigs case, for example, President Kennedy did not question the assumptions made by the CIA and the previous administration, and the result was a disaster. In the Cuban missile crisis, however, the president and his advisors questioned and finally dismissed advice from the military and the hawks in favor of a less provocative and more satisfactory approach. The pressures toward groupthink work in every kind of organization as part of an internal culture with particular assumptions and decision-making processes. General Motors' decision to make the Hummer (while Toyota developed the Prius) perhaps reflects similar dynamics in which obvious questions were not asked and better information was ignored.

It is well documented that pressures such as resource scarcity, drought, extreme heat, and crowding increase social tensions, leading to violence and genocide against minorities. In hard times, authoritarianism flourishes as people prefer order over civility and civil liberties. Building on the earlier work of Theodore Adorno (Adorno et al., 1950), Robert Altemeyer has extended our understanding of the authoritarian personality. It is highly submissive to established authority, inclined to associate only with like-minded persons, believes itself to be highly moral, has little capacity for critical thinking, and is plagued by hypocrisy and double standards (Altemeyer, 1996 and 2004).

From the work of psychologist Erik Erickson (1963) and others, we know that childhood matters. Many personality traits, including willingness to trust, openness to experience, creativity, and problem-solving skills, are formed early in life. A society shortchanges child raising at its peril. We know that exposure to violence early in life can wither the parts of the mind associated with compassion and empathy. Harvard professor Martin Teicher's research shows, for example, that:

> Stress sculpts the brain to exhibit various antisocial, though adaptive, behaviors. Whether it comes in the form of physical, emotional or sexual trauma or through exposure to warfare, famine or pestilence, stress can set off a ripple of hormonal changes that permanently wire a child's brain to cope with a malevolent world. Through this chain of events, violence and abuse pass from generation to generation as well as from one society to the next... once these key brain alterations occur, there may be no going back. (Teicher, 2002, p. 75)

Humanist psychologists like Carl Rogers (1961), Abraham Maslow (1971), and Erich Fromm (1981) have shown that emotion and rationality are not separate things but are so intertwined as to be parts of a larger whole. In the absence of emotion, pure rationality can lead to outcomes such as Auschwitz, while pure emotion without a measure of rationality is ineffective (Damasio, 1994). But there is a strong tendency in Western culture to stifle the expression of emotion only to have it erupt dangerously elsewhere. At some level, we understand the distinction between emotion and rationality to be false. It is emotion that causes a scientist to stay late to check the data one more time. And if we need medical attention or a good lawyer, we aren't likely to seek out professionals with no emotional commitment to health and fairness. We have emotions for good evolutionary reasons, and, as Pascal noted, our hearts guide our rationality, not the other way around. Pascal's point is confirmed by work in neuroscience showing that the emotions influence cognition more than cognition influences emotion (LeDoux, 1996). This may help explain why we are so susceptible to the influence of fear, once a highly adaptive mechanism but one that now threatens the human future.

We know that we succumb to a variety of cognitive traps that undermine our reasoning and the prospect of rational judgments of risk (Ferguson, 2008, pp. 345–346). We are prone to accept information that is close at hand regardless of its relevance. We are inclined to place undue confidence in quantitative risk assessments regardless of validity. We tend to confuse risks associated with known events with the uncertainties of unknown and unknowable probabilities, what risk analyst Nassim Taleb (2008) calls "Black Swans."

Finally, we know that erroneous thinking can sometimes cause us to act in ways that create self-fulfilling prophecies leading to a "reign of error" (Merton, 1968, p. 477). It matters greatly how and how accurately we define ourselves and situations, because we tend to perceive what we assume to be true and act accordingly. Neoclassical economists, for example, define humans as self-maximizing creatures dedicated solely to their own advancement. But this at once purports to be both a description of how humans actually behave and a prescription for how they should behave. Hidden beneath the theory is confusion and conflation of self-interest, which is unavoidable, with selfishness, which is not. This is a basic category mistake that works considerable mischief by justifying individualism at the cost of community.

I think we know as well that the study of mind as practiced from the 18th century to the present has its own limitations and pathologies. Early on it was corseted with assumptions that people are merely machines, that minds and bodies are separate things, and that what can't be counted doesn't count. Modern science, psychology not excepted, was imprinted with faith in progress and the infallibility of human rationality. The problem, however, is not that the authors of Western culture were wrong but rather that we believed them too much for too long. And there were wiser minds all along who knew that the atomistic self of the West was an illusion.

The difference between a future ranging from outright catastrophe to the evolution of a global civilization moving toward justice and sustainability will come down to our capacity to understand ourselves more fully at all levels, ranging from individuals to the deeper and wider currents of mass psychology. Leaders, specifically, will need to understand as never before how to foster the habits of thought and clarity of mind that will enable us to do the things that we must do in order to go through difficult times. Historian and student of leadership James MacGregor Burns distinguishes two types of leadership, transactional and transformational. The former is essentially a broker between competing interests at normal times when the stakes are low (Burns, 2003, p. 24). Transformational leaders, on the other hand, "define public values that embrace the supreme and enduring principles of a people...at testing times when people confront the possibilities—and threat—of great change" (p. 29). By all reckonings the years ahead will be a testing time, calling for both transformational leaders and transformational followers. At all levels, leaders must be master psychologists, empowering and inspiring, not simply ruling, followers. They must help foster the traits necessary to a higher order of human behavior, among which I include gratitude, openness, compassion, generosity, good-heartedness, mercy, tolerance, empathy, humor, courage, and attachment to nature. Listing such things sounds naïve and Pollyannaish, but is less so than might first appear. What do we actually know about the cultivation of such traits?

First, we know that sociability, kindness, and other positive traits are not at all rare: to the contrary, they are common in human behavior. For all of the evidence of human depravity, there are countless cases to the contrary. Villagers in Le Chambon-sur-Lignon, France, for one example, at great risk to themselves, hid Jews from Nazis in World War II (Hallie, 1994). John Rabe, a German citizen living in war-torn Nanking in the 1930s, risked his life to maintain an international safety zone for civilians whose lives were at risk, thereby saving 200,000 Chinese from certain death (Rabe, 2000). Auschwitz survivor Viktor Frankl testifies to the endurance and resilience of prisoners in the midst of unspeakable horrors (2004). In the death camps Tzvetan Todorov found "many more acts of kindness than those recognized by the traditional moral perspective . . . even under the most adverse circumstances imaginable, when men and women are faint with hunger, numb with cold, exhausted, beaten, and humiliated, they still go on performing simple acts of kindness—not everyone and not all the time, but enough to reinforce and even augment our faith in goodness" (1996, pp. 290–291). The lesson, he argues, is that "moral reactions are spontaneous, omnipresent, and eradicable only with the greatest violence" (p. 39). Many war heroes risked their lives for fellow soldiers. There is the everyday heroism of police, firefighters, teachers, and parents who do remarkable things without expecting any tangible reward. And contrary to economic models, there is the everyday kindness of strangers that defies calculations of self-interest.

Good evidence exists, as well, that we have an affinity for life— what E. O. Wilson calls "biophilia." And it would be surprising indeed, after several million years of evolution, were it otherwise. But the study of the ecological foundations of human psychology, beginning with the provocative work of Theodore Roszak (1992), is presently marginalized by mainstream psychology. A few scholars are studying why and how people connect to nature and why that is important for our well-being. Mayer and Frantz, for example, show that the experiential sense of feeling connected to nature is associated not only with greater happiness and meaningfulness in a person's life but also an improved ability to cope with problems (Mayer and Frantz, 2004; Mayer et al., in press)

Biophilia is emerging as an important component of architecture and landscape design (Kellert, Heerwagen, and Mador, 2008). Design that calibrates with our senses by including light, natural materials, white sound, and connection to nature tends to promote learning, accelerate healing in hospitals, increase productivity in the workplace, and likely a great deal more. The design of spaces, landscapes, and streetscapes also affects human behavior in powerful and pervasive ways (Sullivan, 2005; Kaplan and Kaplan, 1989).

The creation of a sustainable society depends on improving the psychological health of individuals and their sense of connectedness to others and nature, but it also depends on improving the capacity of organizations and governments to learn. One of the best examples of corporate learning I know is that of Interface, Inc., the largest U.S. manufacturer of carpet tiles. Founder and CEO Ray Anderson shifted the priorities of the company in the mid-1990s to eliminate waste and use of fossil fuels and is now within reach of that goal. The Interface example is presently the gold standard for corporate learning, and others like Wal-Mart are traveling along the same path.

There are many other examples of decent, creative, and resilient behavior across many domains of experience. They need to be studied, understood, and applied to better advantage in the global effort to build a durable civilization. An urgent challenge for the discipline of psychology, and for students of mind more broadly, is to apply their professional skills to better understand our connections to nature and how to help foster the psychological traits of mind and behavior necessary for a decent future.

It is a challenge as well to the users of psychological research, including advertisers, graphic artists, political consultants, and communications experts, to adopt more stringent codes of conduct that appeal to the angels of our better nature. Abraham

Maslow (1971) traced the development of a full-fledged human being from "infantile self-gratification" through various stages, culminating (for a very few) in transcendence from self. Unfortunately, the theories and empirical data from psychological research are too often applied to manipulate people, aiming to keep them infantilized for commercial or political reasons. We need something akin to the Hippocratic Oath to discipline the application of psychological research, as well as clear standards to guide its use for human development and growth, not exploitation.

In the immediate future we will need the help of psychologists and other social scientists to develop and apply better indicators of human well-being. In 1998, for example, the king of Bhutan stated the objective of using "Gross National Happiness" instead of the standard measure of gross domestic product (Layard, 2005, p. 77). This sounds radical, but only reflects what we already know in our bones: that our well-being, both physical and emotional, grows out of the depth of our connections to each other, to nature, and to our ancestors, and from the faith that we can prevail through the trials of an unknown future.

The evidence from psychology and history, unsurprisingly, tells us that under duress human behavior ranges from very bad to very good. So what makes the difference? One answer lies at the level of cultural narrative—the stories and myths by which we understand larger realities. In Neil Postman's words, such "stories are sufficiently profound and complex to offer explanations of the origins and future of a people; stories that construct ideals, prescribe rules of conduct, specify sources of authority, and, in doing all this, provide a sense of continuity and purpose" (Postman, 1999, p. 101). Whatever the story, as Postman puts it, "human beings cannot live without them. We are burdened with a kind of consciousness that insists on our having a purpose" (p. 101). The narratives that animated the Enlightenment, for example, included the idea of a benign God, the possibility of rational inquiry, the intention to use science to improve people's lives, the faith in progress, the belief in self-governance, and the belief that all men are created equal. The present narrative, at least the commercial version of it, is not so noble, having to do with the promotion of the seven deadly sins of pride, greed, envy, lust, sloth, anger, and gluttony, along with a strong dose of national self-righteousness. But now that we are approaching the edge of a disaster of our own making, what is the right narrative for our time? Frankly, I do not know, but let me suggest three possibilities about the spirit in which we might rewrite our national story.

The first is taken from a friend who recently spent several months as a patient in a cancer ward. During hours of treatment, he witnessed the growth of community among his fellow cancer patients. Once reluctant to say much about themselves, under the new reality of a life-threatening disease they gradually became more talkative and open to thinking about their lives and listening to the experiences of other patients. Living in the shadow of death, they were more open to ideas and people, including some that they formerly regarded as threatening or incomprehensible. They were less prone to arrogance and more sympathetic to the suffering of others. They were less sure of once strongly held convictions and more open to contrary opinions. No longer master of their lives, their schedules, or even their bodies, many achieved a higher level of mastery by letting go of illusions of invulnerability, and in the letting go they reached a more solid ground for hope and the kind of humble but stubborn resilience necessary for beating the odds, or at least for living their final days with grace.

Another possible narrative can be drawn from the experience of people overcoming addiction. Alcoholics Anonymous, for example, offers a 12-step process to overcome addiction that begins with self-awareness, leading to a public confession of the problem, a reshaping of intention, the stabilizing influence of a support group, and a reclaiming of self-mastery to higher ends. The power of this narrative line is in the similarity between substance addiction and its effects and our societal addictions to consumption, entertainment, and energy and their destructive effects on our places, our selves, and our children.

A third narrative comes from the haunting story of the native American Crow Chief Plenty Coups, told by philosopher Jonathan Lear (2006). Under the onslaught of white civilization, the world of the Plains Tribes collapsed, and their accomplishments disappeared along with their culture, sense of purpose, and meaning. At the end of his life Plenty Coups told his story to a trapper, Frank Linderman, saying: "But when the buffalo went away the hearts of my people fell to the ground, and they could not lift them up again. After this nothing happened" (p. 2). Of course many things happened, but without the traditional bearings by which they understood reality or themselves, nothing happened that the Crow people could interpret in a familiar framework. Lear describes Chief Plenty Coups' courageous efforts to respond to the collapse of his civilization with "radical hope," but without the illusion that they could ever re-create the world they had once known. There were others, like Sitting Bull, who pined for vengeance and a return to a past before the juggernaut of American civilization swept across the plains. Likewise, the Ghost Dancers hoped fervently to restore what had been, but Plenty Coups knew that the Crow culture organized around the hunt and warfare would have to become something inconceivably different. The courage necessary to fight had to be transformed into the courage to face and respond creatively and steadfastly to a new reality with "a *traditional* way of going forward" (p. 154). What makes his hope radical, Lear says, "is that it is directed toward a future goodness that transcends the current ability to understand what it is. Radical hope anticipates a good for which those who have the hope as yet lack the appropriate concepts with which to understand it" (p. 104).

In each case, the task of transformational leaders is to help change what is otherwise a disaster at the personal or cultural level into an invitation to openness, catharsis, growth, and renewal—but not to retreat back to the status quo. In each case, the solution requires honesty, introspection, and the admission of hurt and vulnerability. Similarly, the self-induced crisis of planetary destabilization is an invitation for transformational leaders to help us rethink our place in the world and the way we relate to each other and to the larger web of life and radically reconsider our prospects. In each case, the narrative includes the recognition that as some things are ending, other possibilities are beginning. Americans in conditions of climate instability and the end of the era of cheap fossil fuels will witness the end of lots of things, some of which will be of the good-riddance sort, while others will be more painful. Transformational leaders will help us summon honesty and courage to admit that we are the chief culprit in driving the global changes now under way and to discard the belief that by more drilling, mining, economic growth, heroic technology, or military power we can keep the world as it once was. The world in which those were useful or appropriate responses to our problems is disappearing before our eyes.

Do we have grounds for optimism? In the near term, I do not think so. We have yet to attain full awareness of our situation, let alone what will be required of us to deal with it. But I believe as well that the dawn of awareness is close at hand. When we do settle down to work to stabilize climate, good possibilities will take decades or longer to reach the scale of deployment necessary to reverse the accumulation of carbon and defuse other crises. In the mid to longer term, grounds for optimism depend on how rapidly and creatively we make four fundamental changes.

The first necessary change is a radical improvement of societal resilience by reshaping the way we provision ourselves with food, energy, water, and economic support. Resilience implies the capacity to withstand and recover from disturbances, but critical parts of our infrastructure, including the electric grid, energy systems, food system, information technologies, and transportation networks, are highly vulnerable, not just to terrorism but to the cascading effects of breakdowns, accidents, and acts of God. Economist Barry Lynn similarly argues that the same vulnerabilities characterize the global economy that is "ever more interactively complex and tightly coupled" while becoming less redundant and less well managed (2005, p. 234). The issue has a long pedigree.

In 1978, geochemist Harrison Brown proposed a national strategy of resilience that would build "redundancies into the system by endowing the system with more effective means for repairing itself by establishing buffering mechanisms such as improved storage facilities for food and raw materials." His vision included cities that would be self-reliant for food, energy, and materials, in the manner of peasant villages (1978, p. 218, pp. 242–244). Amory and Hunter Lovins' book *Brittle Power* is a blueprint for a resilient energy system, based on nine principles of resilient design that are more broadly applicable as well (1982, pp. 177–213). Yale sociologist Charles Perrow, in his classic 1984 book *Normal Accidents* and more recently in *The Next Catastrophe,* proposes to increase societal resilience by downscaling and decentralizing organizations of all kinds, as well as the electric grid and industrial supply chains (2007, p. 296). But in the absence of any coherent national effort to advance resilience, many citizens are taking matters into their own hands by building local self-reliance for food, energy, and economic support. The movement to build agriculture systems organized on the principles of natural systems, the growth of community-supported farms, the burgeoning Slow Food movement, school gardens, and urban gardens are all promising movements toward resilience (Pollan, 2008). In energy systems, the rapid deployment of wind and solar, even with little government support, similarly reflects the kinds of changes that promote societal resilience and locally based prosperity. But these still isolated and intermittent efforts must be integrated into the broader national effort now under way to improve the resilience, redundancy, and robustness of basic infrastructure and systems.

The second necessary change is a shift in our manner of education that alters both the substance and process of learning, from kindergarten through PhD. The goal is what Robert Jay Lifton and Eric Markusen describe as "a modest yet far-reaching realignment of elements of the self" that extend "the capacity of an individual self for concern, caring, loyalty and even love…to the human species as a whole" (Lifton and Markusen, 1990, p. 259). "Species [awareness]," in their words, "inevitably extends to the habitat of all species, to the Earth and its ecosystem" (p. 275). The problem they've identified is not *in* education but *of* education, and requires a more fundamental transformation of our concept of learning relative to the health of the biosphere.

Michael Crow, president of Arizona State University, describes the problem *of* education in this way: "the academy remains unwilling to fully embrace the multiple ways of thinking, the different disciplinary cultures, orientations, and approaches to solving problems that have arisen through hundreds if not thousands of years of intellectual evolution…Hubris…is a major obstacle to coming to grips with our situation" ("None Dare Call It Hubris," 2007, pp. 3–4). The point is that education has long been a part of the problem, turning out graduates who were clueless about the way the world works as a physical system or why that knowledge was important to their lives and careers, while at the same time promoting knowledge of the sort that has fueled the destruction of ecologies and undermined human prospects.

What would it mean for the academy to deal seriously with the crisis of sustainability, including its underlying causes? Crow's answer, in part, is to "recognize our responsibility to use the knowledge we advance for the good of society" ("American Research Universities," 2007, p. 3). Crow, however, goes further, aiming to restructure the academy as a part of the larger effort to redesign urban communities for sustainability, bringing the considerable intellectual power of the university to bear on local and regional decisions and problems. The 4,100 colleges and universities in the

United States in 2005–2006 had 17.5 million students and 2.7 million faculty and staff, spent $364 billion, and added $28 billion to their endowments (Eagan et al., 2008, p. 8). In other words, schools, colleges, and universities have considerable leverage on our possibilities. Were they to exercise their leadership not only to educate generations of ecologically literate change makers but also to use their buying and investment power to build local and regional resilience, they could greatly speed the transition to a decent future.

The third transition is far more difficult: the reform of our political life. We live amidst the ruins of failed isms. Communism and socialism surely failed, but for different reasons. Capitalism as it is presently practiced, however, is not far behind in the race to oblivion. The first two failed because they promised too much and delivered too little at too high a cost. Global capitalism of the neoliberal variety is failing because it delivers too much to too few far too destructively. With enough historical perspective, the differences among these three systems will seem very small, rather like those minute doctrinal quarrels that fueled religious wars over the centuries. They differ mainly about who owns the means of production, but not a whit about the priority of economic growth. In the meantime, neoconservative devotees in the United States have dismantled much of the capacity for governance in a fit of what Vaclav Havel calls "market madness," which "can be as dangerous as Marxist ideology" (Havel, 1992, p. 66). In fact, ideologues of the extreme right in their neocon phase bear a distinct intellectual and behavioral resemblance to the communists of the Soviet Union. Both cultivated the arts of ruthlessness and manipulation without a flicker of self-doubt about their particular ends justifying their means.

The solutions, long obvious to serious students of democracy, are to end the new gilded age in jackboots by removing money changers from the electoral process once and for all, along with the influence peddlers who descend like locusts on Congress.

Having done that, we might get down to the serious business, in Eric Roston's words, of "weaning civilization from the fuels that enable it, without disrupting civilization . . . the most difficult civic works project ever undertaken—much harder than growing civilization in the first place" (Roston, 2008, p. 187). That will require, among other things, recalibrating governance to the way the world works as a physical system—work begun in the late 1960s and early 1970s by a bipartisan coalition that passed the National Environmental Policy Act, the major pollution control laws, and the Endangered Species Act, among other legislation. We now need to return to that agenda, transcending party affiliations and ideologies of left and right. In the present emergency we will have to act with extraordinary unity and foresight. And the list is long.

We must, in short order, build a world secure by design, restore fairness to the tax system, rebuild democracy, and relearn the civil arts of deliberation and thoughtful civic engagement. We must create an economy fashioned to protect natural capital, rebuild cities, and re-create intercity and light rail transportation systems that were dismantled long ago for the greater convenience and profit of the auto industry. We must devise policies that penalize carbon-based fuels while rewarding efficiency and the use of solar and wind power. We must devise public and private ways to preserve soils, forests, biological diversity, and open spaces. We must restore the capacity of government to ensure fairness, provide equal access to justice, and protect the commons, including the public airwaves now dominated by the merchants of fear, ridicule, and discord. We must create a system that dispenses justice fairly, not merely expand prisons for young males who are disproportionately Black or Latino. We must harness the energy and creativity of all our citizens to build a green economy and solar energy. And we must refashion our neighborhoods, communities, towns, and cities in ways that overcome the "nature deficit disorder" that isolates our children from the natural world (Louv, 2005). That means a society with fewer highways and more bike trails, fewer malls and better schools, less television and more parks, fewer smokestacks and more windmills, fewer gangs and more attentive parents, fewer jobs outsourced to inhuman sweat shops overseas and more permanent, well-paying, green jobs in the local economy. That would, indeed, be not nirvana but a "kinder and gentler society," and that leads to the fourth transition.

On October 2, 2006, a lone gunman entered an Amish schoolhouse near the village of Nickel Mines, Pennsylvania, and opened fire, killing five girls and severely wounding five others. In the aftermath, what was surprising was not that yet another well-armed gunman had snapped and gone on a killing spree, but rather the Amish response. Instead of anger, recrimination, and lawsuits, within hours of the shootings the Amish reached out to the killer's family, offering forgiveness, mercy, and help (Kraybill, Nolt, and Weaver-Zercher, 2007, p. 43). Instead of hatred and revenge, the response was to offer the killer's widow and children friendship and support. At one open-casket funeral, the grandfather of one of the victims admonished the younger children not to "think evil of the man who did this" (p. 45). At the killer's funeral, "About thirty-five or forty Amish came to the burial. They shook [the family's] hands and cried. They embraced Amy [the killer's widow] and the children. There were no grudges, no hard feelings, only forgiveness" (p. 46). The acts of forgiveness were "neither calculated nor random," but rather "emerged from who they were long before" (p. xii). The Amish take the admonition to avoid violence and forgive their transgressors seriously. Amish forgiveness, nonetheless, raises many perplexing questions. Should we forgive those who in cold blood harm others? Should forgiveness extend to those who commit particularly heinous crimes? Should forgiveness extend to persons who show no remorse for their actions? Should Simon Wiesenthal (1997) have forgiven the young, dying Nazi storm trooper who begged for his forgiveness? To raise such questions is to go into a realm in which reason doesn't help much.

For the Amish, forgiveness is "just standard Christian forgiveness," but its Biblical roots are in the practice of *Gelassenheit,* which can be translated as "yieldedness" or "submission" and acceptance of God's will (Kraybill, 100).

The example of the Amish of Nickel Mines is only one example of applied grace. Others can be drawn from other times, cultures, and religious traditions. In Buddhism, for example, compassion is expressed as empathy, "our ability to enter into and, to some extent, share others' suffering" (Dalai Lama, 1999, p. 123). But most religions support the kind of grace shown by the Amish at Nickel Mines and the belief that vengeance is not ours to exact.

In the years of the long emergency there will be much to forgive. We may plausibly expect future wars over water and energy. Millions of climate refugees will cross international borders. If governments fail to act, or fail to act wisely and fairly, violence and economic turmoil could spiral out of control. Civility and kindness will become more difficult in hotter and stormier times. In contrast to the Amish, the larger society has not always nurtured the qualities of neighborliness, compassion, mercy, and forgiveness that will be sorely needed in the long emergency. All of which is to say that we will need not only the right policies and better technologies but what Anita Roddick calls "a revolution in kindness" and a generosity of spirit that allows us to gracefully forgive and to be forgiven (Roddick, 2003). And we can hope that our grandchildren and theirs will one day forgive our callousness and dereliction when we knew what we were doing.

CHAPTER 7

Hope at the End
of Our Tether

Fraudulent hope is one of the greatest malefactors, even enervators, of the human race, concretely genuine hope its most dedicated benefactor.

—Ernst Bloch

Nothing that is worth doing can be achieved in a lifetime; therefore we must be saved by hope. Nothing which is true or beautiful or good makes complete sense in any immediate context of history; therefore we must be saved by faith. Nothing we do, however virtuous, can be accomplished alone; therefore, we are saved by love. No virtuous act is quite as virtuous from the standpoint of our friend or foe as it is from our standpoint. Therefore we must be saved by the final form of love which is forgiveness.

—Reinhold Niebuhr

Ma cleared her throat. "It ain't kin we? It's will we?" she said firmly. "As far as 'kin,' we can't do nothin', not go to California or nothin'; but as far as 'will,' why we'll do what we will."

—John Steinbeck, *Grapes of Wrath*

WE LIKE OPTIMISTIC PEOPLE. THEY ARE FUN, OFTEN funny, and very often capable of doing amazing things otherwise thought to be impossible. Were I stranded on a life raft in the middle of the ocean with the choice of an optimist or a pessimist for a companion, I'd want the optimist, providing he did not have a liking for human flesh. Optimism, however, is often rather like a Yankee fan believing that the team can win the game when it's the bottom of the ninth and they're up by a run with two outs, a two-strike count against a .200 hitter, and Mariano Rivera in his prime on the mound. That fan is optimistic for good reason. Cleveland Indian fans (I am one), on the other hand, believe in salvation by small percentages (if at all) and hope for a hit to get the runner home from second base and tie the game. Optimists know that the odds are in their favor; hope is the faith that things will work out whatever the odds. Hope is a verb with its sleeves rolled up. Hopeful people are actively engaged in defying the odds or changing the odds. Optimism, on the other hand, leans back, puts its feet up, and wears a confident look, knowing that the deck is stacked. "Hope," in Vaclav Havel's words, "is not prognostication. It is an orientation of the spirit, an orientation of the heart; it transcends the world that is immediately experienced, and is anchored somewhere beyond its horizons...Hope, in this deep and powerful sense, is not the same as joy that things are going well,...but, rather, an ability to work for something because it is good" (1991, p. 181).

I know of no purely rational reason for anyone to be optimistic about the human future. How can one be optimistic, for example, about global warming? First, as noted above, it isn't a "warming," but rather a total destabilization of the planet brought on by the behavior of one species: us. Whoever called this "warming" must have worked for the advertising industry or the Northern Siberian Bureau of Economic Development. The Intergovernmental Panel on Climate Change—the thousand-plus scientists who study climate and whose livelihoods depend on authenticity, replicability, data, facts, and logic—put it differently. A hotter world likely means:

- More heat waves and droughts;
- More and larger storms;

- Bigger hurricanes;
- Forest dieback;
- Changing ecosystems;
- More tropical diseases in formerly temperate areas;
- Rising ocean levels—faster than once thought;
- Losing many things nature once did for us;
- Losing things like Vermont maple syrup;
- More and nastier bugs;
- Food shortages due to drought, heat, and more and nastier bugs;
- More death from climate-driven weather events;
- Refugees fleeing floods, rising seas, drought, and expanding deserts;
- International conflicts over energy, food, and water;
- Runaway climate change to some new stable state, most likely without humans if we do not act quickly and effectively.

Some of these changes are inevitable given the volume of heat-trapping gases we've already put into the atmosphere. There is a lag of several decades between the emission of carbon dioxide and other heat-trapping gases and the weather headlines, and still another lag until we experience their full economic and political effects. The sum total of the opinions of climate experts recounted in chapter 1 goes like this:

1. We've already warmed the planet by 0.8°C;
2. We are committed to another ~0.5°C to 1.0°C warming;
3. It's too late to avoid trauma, but;
4. It's *probably* not too late to avoid global catastrophe, which includes the possibility of runaway climate change;
5. There are no easy answers or magic bullet solutions;
6. It is truly a global emergency.

Whether or not item 4 above is correct is anyone's guess, since the level of greenhouse gases is higher than it has been in the past

650,000 years, and quite likely a great deal longer. We are playing a global version of Russian roulette, and no one knows for certain what the safe thresholds of various heat-trapping gases might be. Over the past three decades, scientific certainty about the pace of climate change has had a brief shelf life, but the overall pattern is clear. As scientists learn more, the findings are almost without exception worse than previously thought. In a matter of decades, ocean acidification went from being a concern a century or two hence to being a near-term crisis. The ice sheets of Greenland and Antarctic, once thought to be highly stable, are melting faster than scientists thought possible even a few years ago. The threshold of perceived safety went down from perhaps 560 parts per million CO_2 to 450 ppm, and now James Hanson tells us that it is actually closer to 300 ppm.

Feeling optimistic in these circumstances is like whistling as one walks past the graveyard at midnight. No good case can be made for it, but the sound of whistling sure beats the sound of the rustling in the bushes beside the fence. But whistling doesn't change the probabilities one iota, nor does it much influence any goblins lurking about. Nonetheless, optimistic people calm, reassure, and sometimes motivate us to accomplish a great deal more than we otherwise might. But sometimes optimism misleads, and on occasion badly so. This is where hope enters.

Realistic hope, however, requires us to check our optimism at the door and enter the future without illusions. It requires a level of honesty, self-awareness, and sobriety that is difficult to summon and sustain. I know a great many smart people and many very good people, but I know far fewer people who can handle hard truth gracefully without despairing. In such circumstances it is tempting to seize on anything that distracts us from unpleasant things.

Authentic hope, in other words, is made of sterner stuff than optimism. It must be rooted in the truth as best we can see it, knowing that our vision is always partial. Hope requires the courage to reach farther, dig deeper, confront our limits and those of nature, and work harder. Optimism doesn't have to work very hard, since it is likely to win anyway, but hope has to hustle, scheme, make deals, and strategize. How do we find authentic hope in the face of climate change, the biological holocaust now under way, the spread of global poverty, seemingly unsolvable human conflicts, terrorism, and the void of world leadership adequate to the issues?

Not long ago I was admonished to give a "positive" talk to a gathering of ranchers, natural resource professionals, and university students. Presumably the audience was incapable of coping with the bad news it was assumed that I would otherwise deliver. I gave the talk that I intended to give and the audience survived, but the experience caused me to think more about what we say and what we can say to good effect about the kind of news that many readers of this book reckon with daily.

The view that the public can handle only happy news, nonetheless, rests on a chain of reasoning that goes like this:

- we face problems that are solvable, not dilemmas that can be avoided with foresight but are not solvable, and certainly not losses that are permanent;
- the public, manipulated by advertising and a mendacious media, can't handle much truth, so
- resolution of different values and significant improvement of human behavior otherwise necessary are impossible;
- greed and self-interest are in the driver's seat and always will be, so
- the consumer economy is here to stay, but
- consumers sometimes want greener gadgets, and
- capitalism can supply these at a goodly profit and itself be greened a bit, but not improved otherwise, so
- matters of distribution, poverty, and political power are nonstarters, therefore

- the focus should be on problems solvable at a profit by technology and policy changes;
- significant improvement of politics, policy, and governance are unlikely and probably irrelevant, because
- better design and market adjustments can substitute for governmental regulation and thereby eliminate most of the sources of political controversy.

Masquerading as optimism, this approach is, in fact, pessimistic about our capacity to understand the truth and act nobly. So neither we nor those who presumably lead us talk about limits to growth, unsolvable problems, moral failings, the unequal distribution of wealth within and between generations, emerging dangers, impossibilities, technology gone awry, or necessary sacrifices. "Realism" so diluted requires us to portray climate change as an opportunity to make a great deal of money, which it may be for some, but without saying that it might not be for most, not to mention its connections to other issues, problems, and dilemmas or the possibility that the four horsemen are gaining on us. No American politician is supposed to talk about coming changes in our "lifestyles," a telling and empty word implying fashion, not necessity or conviction.

Ultimately, this approach is condescending to those who are presumed incapable of facing the truth and acting creatively and courageously in dire circumstances. The idea that we may have to give up something in order to stabilize climate is not to be spoken by any national political leader for fear of losing public support. So they reduce the problem to a series of wedges representing various possibilities that would potentially eliminate so many gigatons of carbon without any serious changes in how we live. But there is no proposed wedge called "suck it up," because that is believed to be too much to ask of people who have been consuming way too much, too carelessly, for too long. The "American way of life" is thought to be sacrosanct. In the face of a global emergency, brought on in no small way by the profligate American way of life, few are willing to say otherwise. So we are told to buy hybrid cars, but not asked to walk, bike, or make fewer trips, even at the end of the era of cheap oil. We are asked to buy compact fluorescent light bulbs, but not to turn off our electronic stuff or avoid buying it in the first place. We are admonished to buy green, but seldom asked to buy less or repair what we already have or just do without. We are encouraged to build LEED-rated buildings that are used for maybe ten hours a day for five days a week, but we are not asked to repair existing buildings or told that we cannot build our way out of the mess we've made. We are not told that the consumer way of life will have to be rethought and redesigned to exist within the limits of natural systems. And so we continue to walk north on that southbound train.

And maybe, told that our hindquarters are caught in a wringer, the public would panic or, on the other hand, become so despairing as to stop us from doing what we otherwise would do that could save ourselves from the worst outcomes possible. This is an old and cynical view of human nature that assumes that public order and prosperity requires manipulating people into being dependent and dependable consumers. People who do for themselves make indifferent consumers and are a hazard to both the economy and social stability. This is the kind of reasoning behind the philosophy of Leo Strauss and his secretive followers, who believe that elites have superior knowledge that the bovine masses could never comprehend. Access to such hidden knowledge licenses its possessors to manipulate the public and tell noble lies to achieve higher ends.[1]

Maybe this is true, and maybe gradualism is the right strategy. Perhaps the crisis of climate and those of equity, security, and economic sustainability will yield to the cumulative effects of many small changes without any sacrifice at all. Maybe changes now under way are enough to save us. Maybe small changes will increase the willingness to make larger changes in the future.

188   FARTHER HORIZONS

State-level initiatives in California and Florida, as well as the
Regional Greenhouse Gas Initiative in the Northeastern states,
are changing the politics of climate. With the leadership of Mayor
Daley and his deputy chief of staff, Sadhu Johnston, cities like
Chicago have developed aggressive climate plans (www.chica-
goclimateaction.org). Deployment of wind and solar systems are
growing at 40 percent or more per year, taking us toward a different energy future. A cap and trade bill will sooner or later pass
in Congress, and maybe that will be enough. Maybe we can win
the game of climate roulette at a profit and never have to confront
the nastier realities of global capitalism and inequity, or confront
the ecological and human violence that we've unleashed on the
world.

But I wouldn't bet the Earth on it.

For one thing, the scientific evidence indicates that we have
little or no margin for safety, and none for delay in reducing greenhouse gas levels before we risk triggering runaway change. So call
it prudence, precaution, insurance, common sense, or what you
will, but this ought to be regarded as an emergency like no other.
Having spent any margin of error that we might have had 30
years ago, we now have to respond quickly and effectively, or else.
That's what the drab language of the Fourth Assessment Report
from the Intergovernmental Panel on Climate Change says. What
is being proposed, I think, is still too little, too late—necessary but
not within shouting distance of sufficient. And it is being sold as
"realism" by people who have convinced themselves that they
have to understate the problem in order to appear to be credible.

For another, climate roulette is part of a larger equation of
exploitation of people and nature, violence, inequity, imperialism,
and intergenerational exploitation, the parts of which are inter-
locked. In other words, heat-trapping gases in the atmosphere are
a symptom of something a lot bigger. To deal with the causes of
climate change, we need a more thorough and deeper awareness

of how we got to the brink of destroying the human prospect and much of the planet. It did not happen accidentally but is the logical working out of a set of assumptions, philosophies, worldviews, and unfair power relations that have been evident for a long time. The wars, gulags, ethnic cleansings, militarism, and destruction of forests, wildlife, and oceans throughout the 20th century were earlier symptoms of the problem. We've been playing fast and loose with life for a while now, and it's time to discuss the changes we must make in order to conduct the public business fairly and decently over the long haul.

The upshot is that the forces that have brought us to the brink of climate disaster and biological holocaust and are responsible for the spread of global poverty—the crisis of sustainability—remain mostly invisible and yet also in charge of climate policy. The fact is that climate stability, sustainability, and security are impossible in a world with too much violence, too many weapons, too much unaccountable power, too much stuff for some and too little for others, and a political system that is bought and paid for behind closed doors. Looming climate catastrophe, in other words, is a symptom of a larger disease.

What do I propose? Simply this: that those who purport to lead us, and all of us who are concerned about climate change, environmental quality, and equity, treat the public as intelligent adults who are capable of understanding the truth and acting creatively and courageously in the face of necessity—much as a doctor talking to a patient with a potentially terminal disease. Faced with a life-threatening illness, people more often than not respond heroically. Every day, soldiers, parents, citizens, and strangers do heroic and improbable things in the full knowledge of the price they will pay. Much depends on how and how well people are led. Robert Greenleaf, one of the great students of leadership, puts it this way: "It is part of the enigma of human nature that the 'typical' person—immature, stumbling, inept, lazy—is capable of great dedication and heroism *if* wisely led" (Greenleaf, 1977, p. 21).

Genuine leaders, including those in the media, must summon the people with all of their flaws to a level of extraordinary achievement appropriate to an extraordinarily dangerous time. They must ask people, otherwise highly knowledgeable about the latest foibles of celebrities, to be active citizens again, to know more, think more deeply, take responsibility, participate publicly, and, from time to time sacrifice. Leaders must help the public see the connections between climate, environmental quality, security, energy use, equity, and prosperity. We must relearn how to be creative in adversity. As quaint and naïve as that may sound, people have done it before, and it's worked.

Telling the truth requires leaders at all levels to speak clearly about the causes of our failures that have led us to the brink of disaster. If we fail to treat the underlying causes, no small remedies will save us for long. The problems can in one way or another be traced to the irresponsible exercise of power that has excluded the rights of the poor, the disenfranchised, and every generation after our own. This is in no small way a direct result of money in politics, which has aided and abetted the theft of the public commons, including the airwaves, where deliberate misinformation and distraction of the public is a growth industry. The right of free speech, as Lincoln said in his address to the Cooper Union in 1860, should not be used "to mislead others, who have less access to history, and less leisure to study it." But the rights of capital over the media now trump those of honesty and fair public dialogue, and will continue to do so until the public reasserts its legitimate control over the public commons, including the airwaves.

Transformational leadership in the largest crisis humankind has ever faced means summoning people to a higher vision than that of the affluent consumer society. Consider the well-studied but little-noted gap between the stagnant or falling trend line of American happiness in the last half century and that of rising

GNP. That gap ought to have reinforced the ancient message that, beyond some point, more is not better. If we fail to see a vision of a livable decent future beyond the consumer society, we will never summon the courage, imagination, or wit to do the obvious things to create something better than what is in prospect.

So, what does a carbon neutral society and increasingly sustainable society look like? My list consists of communities with:

Front porches
Public parks
Local businesses
Windmills and solar collectors
Living machines to process waste water
Local farms and better food
More and better woodlots and forests
Summer jobs for kids doing useful things
Local employment
More bike trails
Summer baseball leagues
Community theaters
Better poetry
Neighborhood book discussion groups
Leagues in which no one bowls alone
Better schools
Vibrant and robust downtowns with sidewalk cafes
Great pubs serving microbrews
Fewer freeways, shopping malls, sprawl, and television
More kids playing outdoors
No more wars for oil or access to other peoples' resources.

Nirvana? Hardly! We have a remarkable capacity to screw up good things. But it is still possible to create a future that is a great deal better than what is in prospect. Ironically, what we must do to avert the worst effects of climate change are mostly the same things we would do to build sustainable communities, improve environmental quality, build prosperous economies, and improve the prospects for our children.

I am an educator and earn my keep by perpetuating the quaint belief that if people only knew more we would behave better. Some of what we need to know is new, but most of it is old, very old. On my list of things people ought to know in order to discern the truth are a few technical things like:

1. The laws of thermodynamics imply that economic growth only increases the pace of disorder, the transition from low entropy to high entropy.
2. The basic sciences of biology and ecology—that is, how the world works as a physical system.
3. The fundamentals of ecological carrying capacity, which apply to yeast cells in a wine vat, lemmings, deer, and humans.

But we ought to know, too, about human fallibility, gullibility, and the inescapable problem of ignorance. So I propose that political leaders at all levels, as well as corporate executives, media moguls and reporters, financiers and bean counters, along with all college and university students, read, mull over, and discuss Marlowe's *Dr. Faustus*, Mary Shelley's *Frankenstein*, Melville's *Moby Dick*, and the book of Ecclesiastes as antidotes to the technological fundamentalism of our time. I hope that we would learn how to distinguish those things that we can do from those that we should not do. And the young, in particular, should be taught the many disciplines of applied hope, which include the skills necessary to grow food, build shelter, manage woodlots, make energy from sunlight and wind, develop local enterprises, cook a good meal, use tools skillfully, repair and reuse, and talk sensibly at a public meeting.

And one thing more. Hope, authentic hope, can be found only in our capacity to discern the truth about our situation and ourselves and summon the fortitude to act accordingly. We have it on high authority that the truth will set us free from illusion, greed, and ill will—and, perhaps with a bit of luck, from self-imposed destruction—but that will require a deeper and more fundamental transformation. But exactly what does this mean?

We have come to what Alastair McIntosh calls "a great dying time of evolutionary history" (McIntosh, 2008, p. 191). Some of the traits, skills, and abilities that enabled humankind to survive and eventually to thrive over the millennia are now dangerous to our future. I refer specifically to our fondness for violence. The means of mass destruction are now cheap and easily accessible to nation-states, terrorist organizations, and the merely demented alike. Amplified by conflicts over oil, land, and resources, the likelihood of their use is only a matter of time and circumstance, barring a transformation that seems now to be almost inconceivable.

Throughout history we've tried brute force over and over again, and that is the lamentable story of empires rising and falling. In 1648 the creators of the Westphalian system of sovereign nation-states tried to improve things slightly by creating a few rules to govern interstate anarchy in Europe. The architects of the post–World War II world similarly made incremental improvements by creating international institutions such as the World Bank, the International Monetary Fund, and the United Nations. Nonetheless, war and militarization have a stronger hold on human affairs than ever and threaten, sooner or later, to devour the human prospect.

In the last few centuries we applied the same mind-set to nature. We've bullied, bulldozed, and reengineered her down to the gene, and that got us into more trouble and perplexities than a dozen scientific journals could adequately describe. Some now propose that we manage nature even more intensely—but the same goal with smarter methods will only delay the inevitable. Either way, we are rapidly creating a different Earth, and one we are not going to like. We can quibble about the timing of disaster, but, given our present course, there is no serious argument about its inevitability.

Whether to nature or human affairs, we continue to apply brute force with more powerful and more sophisticated technology and expect different results—a definition, according to some, of insanity. Insanity or not, it is certainly a prescription for the destruction of nature and civilization that is woven into our politics, economies, and culture. The attempt to master nature and to control destiny through force has not worked and will not work, because the world, whether that of nature or that of nations, as Jonathan Schell puts it, is "unconquerable" (Schell, 2003). The reasons are to be found in the mismatch between the human intellect and the complexity of nonlinear systems. No amount of research, thought, or computation can fill that void of ignorance, which is only to acknowledge the limits of human foresight and the inevitability of surprises, unforeseen and unforeseeable results, unintended consequences, paradox, irony, and counterintuitive outcomes. But the limits of human intelligence do not prevent us from discerning something about self-induced messes.

So what kind of messes have we made for ourselves? Some are solvable with enough rationality, money, and effort—like that of powering the world by current renewable energy. However, some situations, like arms races, are not solvable by rational means—although with enough foresight and wisdom they can be avoided or resolved at a higher level. British economist E. F. Schumacher once described the difference between "convergent" and "divergent" problems in much the same terms. In the former, logic tends to converge on a specific answer, while the latter "are refractory to mere logic and discursive reason" and require something akin to a change of heart and perspective (Schumacher, 1977, p. 128). Donella Meadows, in a frequently cited article on the alchemy of change, concluded that of all the possible ways to change social systems, the highest leverage comes not through policies, taxes, numbers or any other item from the usual menu of rational choices, but through change in how we think (Meadows, 1997). Many of our problems can only be worsened by the application of yet more technology, but might be transcended by a change of mind-set.

Two such examples stand astride our age. The first dilemma has to do with age-old addiction to force in human affairs. We don't know exactly how or when violence became the method of choice, or the precise point at which it became wholly counterproductive (Schmookler, 1984). But no tribe or nation that did not prepare for war could survive for long once its neighbors did. And since it makes no sense to have a good army if you don't use it from time to time, preparation for war tended to make its occurrence more likely. If it ever was rational, however, the bloody carnage of the past hundred years should have convinced even the dullest among us that violence within and between societies is ultimately self-defeating and colossally stupid. Violence and threats have always tended to create more of the same—a deadly dance of action and reaction. The development of nuclear and biological weapons and the even more heinous weapons now in development have changed everything... everything but our way of thinking, as Einstein once noted. In an age of terrorism, the scale of potential destruction and the proliferation of small weapons of mass destruction mean that there is no sure means of security, safety, or deterrence anywhere for anyone. The conclusion is inescapable: from now on—whatever the issues—there can be no winners in any violent conflict, only losers. Nonetheless, the world now spends $1.2 trillion each year on weapons and militarism and is, unsurprisingly, less secure than ever. The United States alone spends 46 percent of the total, or $17,000 per second, more than the next 22 nations combined. It maintains over 737 military bases worldwide but is presently losing two wars and has threatened to start a third. Economist Joseph Stiglitz estimates that the total cost of the Iraqi misadventure alone will be $3 trillion. Beyond the economic cost, it will surely leave a legacy of yet more terrorism, violence, despair, and ruin in all of its many guises.

The word "realism" has always been a loaded word. In world politics it is contrasted with "idealism," believed by realists to be the epitome of wooly-headedness. In realist theory, the power realities of interstate politics required military strength and the aggressive protection of the national interest, defined as power. Realists were the architects of empires, world wars, cold wars, arms races, mutual assured destruction, the Vietnam War, and now the fiasco in Iraq. But the difference between idealism and realism has never been as clear as supposed. For example, Hans Morgenthau, one of the preeminent realists of the post–World War II era, once proposed that governments give control of nuclear weapons to "an agency whose powers are commensurate with the worldwide destructive potentials of those weapons" (Joffe, 2007). George Kennan, another post–World War II realist, similarly proposed international measures to prevent both nuclear war and ecological decline— ideas that are anathema to influential neoconservative realists now. Realism is a shifting target, depending on the circumstances and changing realities of the world.

The second dilemma is the impossibility of perpetual economic growth in a finite biosphere. As ecological economists like Herman Daly have said for decades, the economy is a subsystem of the biosphere, not an independent system. The "bottom line," therefore, is set by the laws of entropy and ecology, not by economic theory. The effort to make the economy sustainable by making it smarter and greener is all to the good, but altogether inadequate. It is incrementalism when we need systemic change that begins by changing the goals of the system. Economic growth can and should be smarter, and corporations ought to reduce their environmental impacts, and with a bit of effort and imagination it is possible for most of them to do so. Could we, however, organize all of the complexities of an endlessly growing global economy to fit within the limits of the biosphere in a mostly badly governed world dominated by greed, corruption, corporate competition, and consumerism? The answer is being written in the disappearing forests of Sumatra, in the mountains being flattened in Appalachia, in the 1,000 megawatts per week of new coal plants reportedly being built in China, in the billion dollars of advertisements spent each year to stoke the fires of Western-style consumption, in glitzy shopping malls, in the fantasy world of Dubai, in the temporizing of governments virtually everywhere, and in the corporate pursuit of short-term profit. Progress toward a truly green economy, as Thomas Friedman (2007) notes, is incremental, not transformational, change, and a great deal of it being given lip service thus far is of the smoke and mirrors sort. If we had hundreds of years to make the necessary changes, we might muddle our way to a sustainable economy, but time is the one thing we do not have. If we intend to preserve civilization, the inescapable conclusion is that we need a more fundamental economic transformation, and that means three things that presently appear to be utterly impossible: (1) a change in priorities to facilitate a transition from economic growth (creation of more stuff) to development that genuinely improves the quality of life for everyone, first in wealthy nations and eventually everywhere; (2) the transformation of the consumer economy into one oriented first and foremost to *needs,* not wants; and hardest of all, (3) summoning the compassion and wisdom to fairly distribute wealth, opportunity, and risk. The fact that these three seem wholly inconceivable to most leaders and to most of us indicates the scale of the challenge ahead and the necessity of a different manner of thinking.[2]

Both dilemmas are intertwined at every point. To maintain economic growth, the powerful must have access to the oil and resources of Third World nations, whether those nations like it or not. Global trade, often to the disadvantage of poor nations, requires the use of military forces to patrol the seas, enforce inequities, strike quickly, and maintain pliant governments willing to plunder their own people and lands. The result is animosity that fuels global terrorism and ethnic violence. The power of envy and the desperate search for "a better life" requires the "haves" to build higher fences to keep the poor at bay. Profit and the fear of possible insurrection and worldwide turmoil drives the search for more advanced Star Wars kinds of technology—robot armies, space platforms, and constant electronic surveillance. But our great wealth and many weapons, as Gandhi said, make us cowards, and our fears condone the injustices that underpin our way of life, fueling the hostility that will some day bring it down.

In sum: (1) the time to heal our conflict with the Earth and those between nations and ethnic groups is short; (2) both are dilemmas born in fear, not merely problems; (3) neither can be resolved by applying more of the kind of thinking that created them; (4) the connection between the two is the addiction to violence promoted by much of the media and electronic gaming industry; and (5) neither one can be solved without solving the other.

We are at the end of our tether,[3] and no amount of conventional rationality or smartness is nearly rational enough or smart enough. Climate destabilization, the loss of biological diversity, and the combination of hatred and the proliferation of heinous weaponry are wreaking havoc on our pretensions of control. This is not the time for illusions or evasion; it is time for leadership toward a thorough transformation of our manner of being in the world.

Self-described realists will argue that, however necessary it is, humans are not up to change at the scale and pace I propose— muddling along is the best that we can do. And for those inclined to wager, that is certainly the smart bet. But if that is all that can be said, we have little reason for hope and might best prepare for our demise. On the other hand, not only is transformational change necessary, but it is possible as well. Do we have good reasons to transform the growth economy and transcend the use of force in world politics? Is the public ready for transformation? Is this an opportune time (a "teachable moment" in world history) to do so? Do we have better nonviolent alternatives?

There is a great deal of evidence to suggest a more hopeful view of possibilities than most "realists" are inclined to see. A 2007 BBC poll of attitudes in 21 countries, for example, shows that a majority, including a majority of Americans, are willing to make significant sacrifices to avoid rapid climate change—even though no "leader" has thought to ask them to do so. Can we craft a fair and ecologically sustainable economy that also sustains us spiritually? The present economy has failed miserably on all three counts. As economist Richard Layard puts it, "here we are as a society: no happier than fifty years ago. Yet every group in society is richer" (Layard, 2005, p. 223). Beyond some minimal level, in other words, economic growth advances neither happiness nor well-being. But the outlines of a nonviolent economy are beginning to emerge in the rapid deployment of solar and wind technology, in a growing anticonsumer movement, in the Slow Food movement, and in fields like biomimicry and industrial ecology. In world affairs, the manifest failure of neoconservative realism in the Middle East and elsewhere may have created that teachable moment when we come to our senses and overthrow that outworn and dangerous paradigm for something far more realistic—security for everyone. And at least since Gandhi we have known that there are better means and ends for the conduct of politics.

The transformative idea of nonviolence can no longer be dismissed as an Eastern oddity, a historical aberration, or the height of naïveté. At the end of our tether it is rather the core of a more realistic and practical global realism. It is not *an* option, but the only option left to us. There is no decent future for humankind without transformation of both our manner of relations and our collective relationship with the Earth. Gandhi stands as the preeminent modern theorist and practitioner of the art of nonviolence. His life and thought were grounded in the practice of *ahimsa*, a Sanskrit word that means nonviolence. To denote the practice of *ahimsa* Gandhi coined the word *satyagraha*, which combines the Sanskrit word *satya*, meaning "truth," with *agraha*, meaning "firmness" (Gandhi, 1954, p. 109). Gandhi honed the philosophy of nonviolence into an effective tool of change in India as Martin Luther King Jr. later did in the United States, but we've never known what to do with people like Gandhi and King. On one hand we occasionally pay them lip service in public speeches and name holidays in their honor, but on the other hand we ignore what they had to say about how we live and how we conduct the public business. The time has come to pay closer attention to what they said and did, and to fathom what that means for us now and in the long emergency ahead.

The beginning of a more realistic realism is in the recognition that violence of any sort is a sure path to ruin on all levels and that the practice of nonviolence is a viable alternative—indeed, our only alternative to collective suicide. But that implies changing a great deal that we presently take for granted, beginning with the belief in an unmovable and implacably evil enemy. Richard Gregg, an associate of Gandhi, for example, said that the goal of a practitioner of nonviolence

> is not to injure, or to crush and humiliate his opponent, or to "break his will"... [but] to convert the opponent, to change his understanding and his sense of values so that he will join whole-heartedly [to seek] a settlement truly amicable and truly satisfying to both sides (Gregg, 1971, 51).

As with war, the practice of nonviolence requires training, discipline, self-denial, strategy, courage, stamina, and heroism. Its aim is not to defeat but to convert and thereby resolve the particulars of conflict at a higher level. For Gandhi, living nonviolently required its practitioners first to transcend animosity and hatred to reach a higher level of being in "self-restraint, unselfishness, patience, gentleness" (Gandhi, 1962, p. 326). The aim is not to win a conflict but to change the mind-set that leads to conflict, and ultimately to form a "broad human movement which is seeking not merely the end of war but [the end of] our equally non-pacifist civilization."

In Gandhi's words, "true *ahimsa* should mean a complete freedom from ill will and anger and hate and an overwhelming love for all" (p. 207).

Gandhi applied the same logic to the industrial world of his day, regarding it as a "curse . . . depend[ing] entirely on [the] capacity to exploit" (p. 287). Its future, he thought, was "dark" not only because it engendered conflict between peoples but because it cultivated "an infinite multiplicity of human wants . . . [arising from] want of a living faith in a future state, and therefore also in Divinity" (p. 289).

The philosophy, strategy, and tactics of nonviolence have been updated to our own time and situation by many scholars, including Anders Boserup and Andrew Mack (1975), Richard Falk and Saul Mendlovitz (World Order Models Project), Michael Shuman and Hal Harvey (1993), Gene Sharp (1973, 2005), and the Dalai Lama (1999). We do not lack for examples, precedents, alternatives, and better ideas than those now regnant; we lack the leadership to move. It is time—long past time—to take the next steps in rethinking and remodeling our economy and foreign policies to fit a higher view of the human potential. With clear vision, the first steps will be the hardest of all, because the impediment is not intellectual but something else that lies deeper in our psyche. Over the millennia violence became an addiction of sorts. Most of our heroes are violent men. Our national holidays mostly celebrate violence in our past. Most of our proudest scientific achievements have to do with the violent domination of nature. There is something in us that seems to need enemies even if, sometimes, they have to be conjured up. And to that end we have built massive institutions to plan and fight wars, giant corporations to supply the equipment for war, and a compliant media to sell us war as a patriotic necessity. In the process, we have made economies and societies dependent on arms makers and merchants of death and changed how we think and how we talk. We often speak violently and think in metaphors of combat and violence, so we "kill time," "make a killing" in the market, or wage futile "wars" on drugs, poverty, and terrorism. Worse, our children are being schooled to think violently by electronic games, television, and movies. We have made no comparable effort to build institutions for the study and propagation of peace and conflict resolution or to cultivate the daily habits of peace. We have barely begun to imagine the possibility of a nonviolent economy in which no one is permitted to profit from war or violence in any form. And so it is surprising that we are continually surprised when our collective obsession with violence manifests yet again in violence down the street or in some distant place.

The transformation to a nonviolent world will require courageous champions at all levels—public officials, teachers, communicators, philanthropists, artists, statespersons, philosophers, and corporate executives. But in democratic societies it will most likely be driven by ordinary people who realize that we are all at the end of our tether and it is time to do something a great deal smarter and more decent. And "somebody must begin it." The next step is to begin to rid ourselves of the most heinous weapons in our bloated arsenals. But that requires, in the words of the former commander of the U.S. nuclear forces, George Lee Butler, understanding "the monstrous effects of [nuclear] weapons...and the horrific prospect of a world seething with enmities, armed to the teeth with nuclear weapons, and hostage to maniacal leaders strongly disposed toward their use" (Butler, 1996). The course is clear: we have "to rid ourselves of the attitudes, and the postures, the policies, and the practices that we became so accustomed to as routine" (Smith, 1997, p. 45). The goal is not to control nuclear weapons, it is to rid ourselves of them.

At the end of our tether we must imagine the unimaginable: a world rid of nuclear weapons and a world powered by sunlight, safe from the possibility of catastrophic climate change. Utopia? Hardly. But those are the only realistic options we have.

CHAPTER 8

The Upshot: What
Is to Be Done?

*The question of whether technology, politics and economic muscle
can sort out the problem is the small question. The big question is
about sorting out the human condition. It is the question of how we
can deepen our humanity to cope with possible waves of war, famine,
disease, and refugees....*

—Alastair McIntosh

A S I WRITE, THE PRESIDENT-ELECT AND HIS ADVISORS
are pondering what to do about climate change
amidst the largest and deepest economic crisis since the Great
Depression of the 1930s. Their first round of decisions will have
been made by the time you read this book. But whatever policy
emerges in the form of cap and trade legislation, taxation, and
new regulations on carbon, they are only the first steps, and they
will quickly prove to be inadequate to deal with a deteriorating biophysical situation. Emerging climate realities will drive
this or the next president, probably sooner rather than later, to
more comprehensive measures—as a matter of national and global
survival. The problem for President Obama presently is that we
are running two deficits with very different time scales, dynamics, and politics. The first, which gets most of our attention, is
short-term and has to do with money, credit, and how we create

and account for wealth, which is to say a matter of economics. However difficult, it is probably repairable in a matter of a few years. The second is ecological. It is permanent, in significant ways irreparable, and potentially fatal to civilization. The economy, as Herman Daly has pointed out for decades, is a subsystem of the biosphere, not the other way around. Accordingly, there are short-term solutions to the first deficit that might work for a while, but they will not restore longer-term ecological solvency and will likely make it worse. The fact is that climate destabilization is a steadily—perhaps rapidly—worsening condition with which we will have to contend for a long time to come. University of Chicago geophysicist David Archer puts it this way:

> a 2°C warming of the global average is often considered to be a sort of danger limit benchmark. Two degrees C was chosen as a value to at least talk about, because it would be warmer than the Earth has been in millions of years. Because of the long lifetime of CO_2 in the atmosphere, 2°C of warming at the atmospheric CO_2 peak would settle down to a bit less than 1°C, and remain so for thousands of years (Archer, 2009, pp. 146–147).

But if the record of earlier climate conditions holds true in the future, it also means, among other things, a 10-meter sea level rise as well as warmer temperatures for thousands of years. Climate destabilization, in short, is not a solvable problem in a time span meaningful for us. But we do have some control over the eventual size of climatic impacts we've initiated if we reduce emissions of CO_2 and other anthropogenic heat-trapping gases to virtually zero within a matter of decades. Assuming that we are success-ful, by the year 2050, say, we will not have forestalled most of the changes now just beginning, but we will have contained the scope, scale, and duration of the destabilization and created the foundation for a future better than that in prospect.

There is no historical precedent, however, for what we must do if we are to endure. Our biology, and specifically the way we perceive threats, was honed over the ages to respond to direct physical threats posed by predators animal or human. It did not equip us very well to perceive and respond to threats measured in parts per billion that play out over decades, centuries, and millennia. We respond, as noted above, with alacrity to threats that are big, fast, and hairy, and not so quickly or ingeniously to those that are slow, small, subtle, and self-generated. Our understanding of economics was developed in the industrial age and imperfectly accounts for the damage caused to ecosystems and the biosphere, and not at all for the destabilization of climate. Had it been otherwise, we would have known that we were not nearly as rich as we presumed ourselves to be and not nearly as invulnerable as we thought. Our politics are a product of the European Enlightenment and rest on the belief in progress and human improvement, which we now know are not as simple or as unambiguous as we once thought. The political forms of democracy reflect a bedrock commitment to individual rights but exclude the rights of other species and generations unborn. And it is in the political realm that we must find the necessary leverage to begin the considerable task of escaping the trap we've set for ourselves.

The challenge before the president and his successors, accordingly, is first and foremost political, not economic. Our situation calls for the transformation of governance and politics in ways that are somewhat comparable to that in U.S. history between the years of 1776 and 1800. In that time Americans forged the case for independence, fought a revolutionary war, crafted a distinctive political philosophy, established an enduring Constitution, created a nation, organized the first modern democratic government, and invented political parties to make the machinery of governance and democracy work tolerably well. Despite its imperfections regarding slavery and inclusiveness, it stands nonetheless as a stunning historical achievement. The task now is no less daunting, and even more crucial to our prospects. We need a systematic calibration of governance with how the world works as a physical system. Theories of laissez-faire, however useful for short-term wealth creation, have proved to be ecologically ruinous. Henry David Thoreau in our circumstances would have asked what good is a growing economy if you don't have a decent planet to put it on.

Few have even begun to reckon with changes of this magnitude; instead, we place our faith in better technology and incremental changes at the margin of the status quo, hoping to keep everything else as it comfortably is. There is much to be said for better technology and particularly for measured policy changes and doing things piecemeal, mostly because we are often ignorant of the side effects of our actions. Revolutions generally have a dismal history. But in the age of consequences, we have no real choice but to transform our conduct of the public business in at least three ways. First, and most fundamental, as a matter of public policy we must quickly stabilize and then reduce carbon emissions. To do so will require policy changes that put an accurate price on carbon-based fuels and create the incentives necessary to deploy energy efficiency and renewable energy technologies here and around the world on an emergency basis. Success in this effort requires that the president and his successors regard climate policy as the linchpin connecting other issues of economy, security, environment, and equity as parts of a comprehensive system of policies governing energy use and economic development. The details of such a policy were recommended to President Obama's transition team by the Presidential Climate Action Project (www.climateactionproject. com) immediately after the election of 2008, and many of the recommendations subsequently appeared in the president's climate policy. Beyond the policy details, the president will need to establish some mechanism by which to reliably coordinate national policies across federal and state agencies whose missions often conflict with the overriding goal of reducing carbon emissions.

Second, the president must launch a public process to consider long-term changes in our systems of governance, politics, and law. The goal is to create practical recommendations that enable us to anticipate and surmount the challenges ahead and ensure, as much as is humanly possible, that we never again stumble to the brink of global disaster. To that end I propose the appointment of a broadly based presidential commission to consider changes in governance and politics, including the necessity of a second constitutional convention. Neither idea is new. Presidential commissions have long been used as a way to engage thoughtful and distinguished persons in the task of rethinking various aspects of public policy and governance. The Ash Council, for one, laid the groundwork for what eventually became the U.S. Environmental Protection Agency. And the idea of a new constitutional convention has been proposed by legal scholars as diverse as Sanford Levinson and Larry Sabato, among many others (Levinson, 2006, p. 173; Sabato, 2007, pp. 198–220). In Sabato's words, the founders:

> had risked life, limb, fortune, and birthright to revolt against their mother country, determined to stand on principle...But they might also have been surprised and disappointed that future generations of Americans would be unable to duplicate their daring and match their creativity when presented with new challenges. (pp. 199–200)

Facing challenges that dwarf any that the founders could have imagined, we should be at least as bold and farsighted as they were. Whether a presidential commission would propose to reform governance by legislation, amendments to the Constitution, a full-scale constitutional convention, or some combination of measures, their charge would be to reform our system of governance to improve democracy and promote deliberation in ways that soon produce wise and well-crafted public policies that accord with ecological realities. Beyond proposals by experts like Levinson, Sabato, and Robert Dahl that aim to make our politics more democratic and efficient, I propose that the Constitution be amended to protect the rights of posterity to life, liberty, and property. The people of Ecuador went still farther, changing their constitution in September 2008 to acknowledge the rights of nature and permit their people to sue on behalf of ecosystems, trees, rivers, and mountains,[1] an idea that owes a great deal to Aldo Leopold's 1949 essay on "The Land Ethic" and to Christopher Stone's classic article in 1972 in the *Southern California Law Review*, "Should Trees Have Standing?" (Stone, 1974). What first appears as "a bit unthinkable" in Stone's words, however, is yet another step in our understanding of rights and obligations due some other person, or in this case, an entity, the web of life.[2] And not once in our history has the extension of rights caused the republic to tremble. To the contrary, it has always opened new vistas and greater possibilities, with one potentially fatal exception.

That exception is the rights of personhood presumed granted to corporations by the U.S. Supreme Court in the *Santa Clara County v. Southern Pacific Railroad* decision of 1886. Whether the Court actually made such a grant or not, it is long past time to rein in the power of corporations, for reasons that are patently obvious. "The only legitimate reason for a government to issue a corporate charter," in economist David Korten's words, "is to serve a well-defined public purpose under strict rules of public accountability" (Korten, 2007). That some corporations have got the new religion on energy efficiency or greening their operations or carbon-trading schemes pales beside the fact that none is capable of "voluntarily sacrificing profits to a larger public good," in Korten's words. And with very few exceptions they are incapable of helping us to reduce consumption, promoting public health, increasing equality, cleaning up the airwaves, or restoring a genuine democracy. It is time for this archaic institution to go the way of monarchy and for us to create better and more accountable ways to provision ourselves.

The presidential commission will need to carefully consider other bold ideas. Peter Barnes, for example, has proposed the creation of an Earth Atmospheric Trust based on the recognition that the atmosphere is a public commons (Barnes, 2006; Barnes et al., 2008). Use of the commons as a depository for greenhouse gases would be auctioned, with the proceeds going to a quasi-independent agency and at least partially redistributed back to the public, the owners of the commons. As the cap for emissions was lowered, the Trust would generate increasing revenues to help the public pay for the transition. There are other ideas to better harness and coordinate science with federal policy. One such proposal is to create an "Earth Systems Science Agency" by combining the National Oceanic and Atmospheric Agency and the U.S. Geological Survey to better collaborate with NASA. The new agency would be "an independent federal agency... with direct access to the Congress and Executive Office of the President, including the Office of Science and Technology Policy and the Office of Management and Budget" (Schaefer et al., 2008, p. 45). The larger goal is to better align earth systems science with the creation and administration of public policy at the highest level as rapidly as possible.

Further, I propose the creation of a council of elders to advise the president, Congress, and the nation on matters of long-term significance relating to climate.[3] The group would be appointed by the president with the advice of the National Academy of Sciences, the American Association for the Advancement of Science, and the American Bar Association, as well as civic, religious, academic, business, philanthropic, and educational groups. It would consist of persons who are distinguished by their accomplishments, wisdom, integrity, and record of public service, not just by their wealth. Their role would be, as the Quakers have it, to speak truth to power, publicly, powerfully, and persistently. The Council of Elders would be given the resources necessary to educate, communicate, commission research, issue annual reports, convene gatherings, engage the global community, and serve as the voice of the powerless, including posterity. Perhaps one day it could merge with a similar body summoned by Virgin Airlines owner Richard Branson and including Jimmy Carter, Vaclav Havel, Nelson Mandela, Beatrice Robinson, and Desmond Tutu into a Global Council of Elders. But at any scale the point is the same: in difficult times ahead we will need to hear the voices of the wisest among us to guide, cajole, admonish, and inspire us along the journey ahead. And those who govern will need their counsel, steadiness, and vision.

Beyond the details of policy and a reformed governing system, the president must resuscitate the role of the president as educator-in-chief using the office in the way Theodore Roosevelt once described as a "bully pulpit." Americans will need to learn a great deal about climate and environmental science in a short time. By skillful communications and the use of the powers of the federal government, the president can help to raise public understanding about climate science to levels necessary to create a constituency for the long haul. The president and others in leadership positions will need as well to build the case for:

- federally financed elections to remove money from the electoral process;
- reforming the Federal Communications Commission to restore the "fair and balanced" standards for use of the public airwaves;
- ending the revolving door between government service and private lobbying;
- desubsidizing coal, oil, gas, and nuclear power;
- reducing the Pentagon budget by, say, half to accord with a more modest U.S. world presence in the world and a smarter strategy that aims for security by design for everyone not brute force to protect corporations; and

- making more radical changes that might someday lead a more civilized America to confiscate 100 percent of the profits from making weapons.

As educator-in-chief, the president must help to rebuild our civic intelligence, emphasizing why fairness and decency are fundamental to prosperity and our well-being lest under the duress of hard times we forget who we are. The president must also help to extend our notions of citizenship to include our role as members in the wider community of life and knowledge of why being good citizens on both counts is the bedrock for any durable civilization. The president must help us understand the ties that bind us together and extend our sight to a farther horizon.

The challenge of transformative leadership in the age of consequences, however, does not fall only to the president and those in Washington. Far from it! The greater part of the work will be done—as it always has been—by those in leadership positions in nonprofit organizations, education, philanthropy, media, churches, business, labor, health care, research centers, civic organizations, mayors, governors, state legislators... virtually all of us. It is mandatory that we all contribute to the effort to minimize and then eliminate carbon emissions, deploy solar technologies, make the transition to a post-carbon economy, reengage the international community, and come to regard ourselves as trustees for future generations. This is a paradigm shift like no other. It is what philosopher Thomas Berry calls our "Great Work." Like that of earlier times, it will be costly and difficult, but far less so than not doing it at all.

From nearly a half century of work in sustainable and natural systems agriculture, urban design, biomimicry, ecological engineering, green building, biophilic design, solar and wind technology, regenerative forestry, holistic resource management, waste cycling, and ecological restoration, we have the intellectual capital and practical experience necessary to remake the human presence on the Earth. From intrepid social examples such as those in Kerala, Curitiba, Saul Alinsky's community organizing in Chicago, and the Mondragón Cooperative in Spain, we know how to build locally based economies that use local resources and local talents to the benefit of local people (McKibben, 1995). Thanks to great educators like John Dewey, Maria Montessori, J. Glenn Gray, Alfred North Whitehead, and Chet Bowers, we have a grasp of the changes in teaching and mind-set necessary to make the transition. And from the most prescient among us, like Wendell Berry, Ivan Illich, and Donella Meadows, we know that fast is sometimes slow, more is sometimes less, growth is sometimes ruinous, and altruism is always the highest form of self-interest. This is to say that we are ready to transform our lives, culture, and prospects, and the time is now!

What does this mean on Main Street? I will end on a personal note. I live in a small Midwestern city powered mostly by coal with a struggling downtown threatened by nearby megamalls. The city is roughly a microcosm of the United States in terms of income distribution, ethnicity, and public problems. Run our likely history fast forward, say, 20 years or more and the town would be in disrepair and seriously impoverished. To avoid that scenario, a group of concerned citizens have recently banded together to create another story. They include the president of Oberlin College, the city manager, the superintendent of schools, the director of the municipal utility, the current and former presidents of the City Council, and many others.

The task before us requires solving four problems. The first is to create a practical vision of post-carbon prosperity. Can we make the transition from coal to efficiency and renewable energy in a way that lays the foundation for a sustainable economy? The second challenge is to develop the financial means to pay for the transition, including the capital costs to implement energy efficiency and to build the new energy system. The third challenge is that of actually building an alternative energy infrastructure in

Oberlin, which means expanding existing businesses or building new ones. The fourth is to structure private choices so that people have a clear incentive to choose efficiency and renewables over inefficiency and fossil fuels and to buy more locally made or grown products.

In 2007, with outside support, the college launched two studies to help clarify our basic energy options. The first, by a Massachusetts energy firm, examined smart ways by which the city could improve efficiency and switch to renewable energy and thereby avoid joining in a risky, long-term commitment to a 1,000-megawatt coal plant (without the means to sequester carbon) proposed by AMP-Ohio. The second study, specifically on college energy use, examined options for eliminating our coal-fired plant and radically improving energy efficiency to levels now technologically possible and economically profitable. We now have a factual basis on which to build a farsighted energy policy for both the city and the college.[4] The college commissioned a third study to explore the feasibility of developing a new green, zero-discharge, carbon-neutral arts block on the east side of the town square, including a substantial upgrade of a performing arts center and a new green hotel.

What might that future look like ten years from now? Imagine, first, picking up an Oberlin phone book or going online and finding perhaps five new companies offering energy services, efficiency upgrades, and solar installations. Imagine a city economy that includes a hundred or more well-paying green energy jobs filled with highly trained young people from Oberlin High School, the vocational school nearby, and the college. Imagine local businesses using a third of the energy they now use but with better lighting and better indoor comfort at a fraction of the cost, with the savings forming the basis for expanded services and profits. Imagine a city that is sprouting photovoltaic (solar electric) systems on rooftops, installed and maintained by local entrepreneurs. Imagine the local utility (Oberlin Municipal Power and

Light) becoming a national leader in improving local efficiency (what is called "demand-side management") while actually lowering energy bills for residents. Imagine the possibility of a new four-star, LEED platinum hotel, conference center, restaurant, and perhaps culinary school as the keystone of a new carbon-neutral, zero-discharge downtown arts district that features great live performances in a new theater and a jazz club featuring student artists from the Oberlin Conservatory of Music. Imagine a revitalized downtown bustling 24 hours a day with residents, shoppers, students, artists, and visitors who came to experience the buzz of the best small town in the United States that is also the first working model of post–fossil fuel prosperity.

Imagine traveling just outside the city into New Russia township, where dozens of farms form a green belt around the city. In the summer they employ Oberlin teens, providing useful work and training in the practice of sustainable agriculture. Local farms flourish by supplying the college dining service, local restaurants, and the public with organically grown fresh foods. Beyond the green belt there is another forested belt of 10,000 acres that profitably sequesters carbon and provides the basis for a thriving wood products business. Imagine a resilient town economy buffered to a great extent from larger economic problems because it is supplied locally with biofuels, electricity from sunshine and wind, and a large portion of its food. Imagine Oberlin leading in the deployment of new technologies just coming into the market, like plug-in hybrid cars, solar electric systems, and advanced wastewater treatment systems. Imagine hundreds of Oberlin students, equipped with skills, aptitudes, and imaginations fostered in the remaking of the town and the college spreading the revolution across the United States and the world.

Imagine a town, churches, college, and local businesses united in the effort to create the first model of post-carbon prosperity in the United States, at a scale large enough to be nationally instructive but small enough to be both manageable and flexible.

Imagine that model spreading around the United States, cross-fertilizing with hundreds of other examples elsewhere in large cities like Chicago and Seattle, urban neighborhoods, and small towns. If, for a moment, you get very quiet...you can feel the transformation going on in neighborhoods, town, and cities all over the United States. It has grown into a worldwide movement that rejects the idea that we are fated to end the human experiment with a bang or a whimper on a scorched and barren Earth. It is the sound of humankind growing to a fuller stature—a transformation just in time.

Postscript: A Disclosure

MOUNT ST. HELENS ERUPTED ON MAY 18, 1980. AS THE cloud of volcanic debris passed over the Midwest, it triggered thunderstorms and heavy rains. But that was the last measurable rainfall until early November in the Meadowcreek Valley, in the Boston Mountains of northern Arkansas. The summer of 1980 turned into the hottest and driest summer recorded up to that point in Arkansas history. For the next two months the high temperatures in the Meadowcreek Valley went over 100°F on 59 days. The highest recording was 111 degrees.

As operators of a small farm and sawmill, my brother and I worked in the heat every day. The only time in my life that I thought for nearly certain that I would die of heatstroke began at 5:30 in the morning, working to widen a rough logging road on the west ridge above the valley, called Pinnacle Point, and ended in the early evening after we'd harvested and bagged 40 acres of wheat. Due to a lack of forethought, we had only 120-pound, not 80-pound, burlap bags, which I came to seriously regret in the late afternoon as we loaded them onto the trailer. The temperature on that particular day reached 108°, and only fools tempt fate in such heat and humidity. We qualified. Our only salvation was in the cool waters of a deep swimming hole after work. By late July, however, the swimming hole had disappeared, and Meadowcreek was mostly bone dry. Throughout the Ozarks that summer, springs never known to go dry disappeared.

As the summer days ticked by, the changes in the land, vegetation, animals, and people became daily more evident. Around the limestone bluffs that rimmed the valley, the red oaks turned brown by early August. The smell of smoke from dozens and then hundreds of fires was constantly in the air. One of our cows died, most likely from heatstroke. Twice I saw birds fall from the sky, apparently from heat-induced heart failure. Rattlesnakes seeking water migrated from the ridge tops to the valley floor, leaving serpentine trails in the thick dust as they crossed dirt roads. People became lethargic, and a few became violent. Heat madness was said to be the defense in a shooting nearby.

Heat has a particular smell to it, not altogether unpleasant. You can feel heat in your bones. Exposed day after day to high temperatures and humidity, with little relief at night and without the luxury of air-conditioning, the body's core temperature elevates. Heat can scour your mind, leaving only an obsession with coolness and water. If you work outdoors in extreme heat, you learn to linger in the shade, avoid midday sun when possible, move slowly, and drink lots of water. But if you work indoors under the anesthesia of air-conditioning, it is possible to avoid the experience of heat or drought or the changes in the land. On a day that summer on which some rain was forecast, I heard a DJ on a Little Rock radio station complain that he might miss his golf game. I laughed, but not entirely in mirth. By late September temperatures were cooler, but the rains held off for another month. The first measurable rainfall we had in the valley came at the end of October and felt like manna from heaven. The smell of rain on parched ground is still about the sweetest smell I know.

There were other hot and dry summers in the 1980s, notably that of 1988. On a flight from Memphis to Little Rock I recall seeing a barge stranded in the Mississippi. The river was so diminished that it looked as if a person could jump from the barge to either Tennessee or Arkansas without getting his feet wet. Water to cool power plants stressed by high demands for air-conditioning was in short supply. Driving across the Midwest in August, I read newspaper stories of shootings over trivial arguments in which heat was said to be a factor. Scientist James Hansen testified to Congress that summer, saying that the nation was seeing the first tangible evidence of climate change. Climate skeptics greeted his testimony with scorn, but fewer do today.

No one can say with certainty that the summer of 1980, or that of 1988, or the recent droughts in the Southeast or Southwest, or Katrina, or the floods in Iowa in 2008, or any number of other weather events are the result of anthropogenic climate change. But the odds that they are rise with each increment of temperature increase, and they are certainly consistent with what can be expected in years to come.

After the summer of 1980, climate change was important to me, not because I'd thought a great deal about it in an air-conditioned office but because I had first felt it viscerally and somatically. My interest did not begin with any abstract intellectual process or deep thinking but rather with the felt experience of the thing, or what the thing will be like. That summer is recorded both mentally and bodily in memories of extreme heat with no respite.

In the summer of 1988 our organization, the Meadowcreek Project, Inc., sponsored the first conference ever on the future of the banking industry in a hotter world. Then-Governor Bill Clinton cosponsored the event, which included bankers from Arkansas, Louisiana, Texas, and Oklahoma and climate scientists such as Stephen Schneider and George Woodwell and energy expert Amory Lovins. The point of the event was to advocate changes in banking practices to minimize climate effects and encourage lenders to recognize their self-interest in avoiding loans for energy inefficient projects. One of the bankers, Bill Bowen, said that "if half of what I've heard is correct, what I'm doing is criminal." I responded by saying something to the effect that more than half of what he'd heard was correct. That meeting was 20 years ahead of its time.

After the summer of 1980, the prospect of climate change had a hold on my attention that has grown as the evidence, both scientific and anecdotal, has mounted. Both the climate system and human systems are nonlinear, which is to say that both are subject to rapid and unpredictable changes that can spiral out of control with small provocations at certain times and places. We should have the intelligence and courage to dispel any lingering belief that we can turn up the thermostat of the Earth and assume that nothing else will change. Beyond some unknown point, lots of other things will change, including our own behavior and capacities. We will not be the same people at a consistent daily high of 110° as we are at 85°. Extreme duress will cause governments and corporations to become more erratic. Under extreme heat and drought, wildlife disappears and ecologies wither. Some of our technology will not work dependably in extreme heat. Airport runways and highways can buckle. Steel can warp and bend. Cooling water for power plants will dry up. Air and water pollution will be more concentrated. The summer of 1980 was a small, very small, preview of a world we should avoid—and still just might.

Notes

Preface

1. From different perspectives and for different reasons, both Bill Joy (2000) and Ray Kurzweil (2005) arrive at the conclusion that humans are likely to lose whatever control we have over our own future. Joy takes no particular joy in this fact. Kurzweil, like many in the field of artificial intelligence, is strangely euphoric.

2. The apt phrase is the title of James Howard Kunstler's 2005 book. My use of it, however, is intended more broadly to include problems of climate destabilization, the end of the era of cheap oil, ecological degradation, and related problems. John McHale, similarly, once described the future as a "crisis of crises," implying the same convergence of problems, dilemmas, and systems breakdowns.

3. Solomon et al. (2009) estimate that changes in surface temperature, rainfall, and sea level are irreversible for more than 1,000 years after carbon dioxide emissions are completely stopped.

4. See www.climateactionproject.com and Becker (2008).

Introduction

1. See also Rees (2008), pp. 41–44, and Smil (*Global Catastrophes*, 2008), who after surveying things is "deliberately agnostic about civilization's fortunes" while acknowledging that "none of us knows which threats and concerns will soon be forgotten and which will become tragic realities" (p. 251).

2. "Carbon cycle feedbacks" refers to positive feedbacks that occur as temperatures rise above their normal range. The most commonly cited are those presently occurring in the Arctic in which surfaces covered with ice with high albedo that reflects sunlight back into space melt and are replaced with dark land and water surfaces that absorb solar radiation, thereby adding to further warming. There are many, many such mechanisms that are sensitive to slight changes that initiate accelerating changes. See Woodwell (1995) and Archer (2009), pp. 125–136.

3. For the sake of clarity, I assume that scenarios presented by Lester Brown (2008) as the "great mobilization," Krupp (2008), and Thomas Friedman (2008) more or less describe what will come to be true. In other words, I will be as optimistic as the science and rationality permit one to be, but no more.

4. See, for example, Holdren, "Science and Technology" (2008): "A 2007 report for the UN Commission on Sustainable Development...concluded that the chances of a 'tipping point' into unmanageable degrees of climatic change increase steeply once the global average surface temperature exceeds 2°C above the pre-industrial level.... Having a better-than-even chance of doing this means stabilizing atmospheric concentrations of greenhouse gases and particles at the equivalent of no more than 450 to 400 parts per million by volume of CO_2." See also Gulledge (2008).

5. The evidence that climate change could be more rapid and more severe than commonly thought is summarized in Pittock (2008); also Alley (2004).

6. For a good summary of the science about changes at given levels of climate forcing see Lynas (2007) and Romm (2007), pp. 27–95.

7. See also Porritt (2006), p. 219.

Chapter 1

1. See also Robert Nelson's thoughtful critique (Nelson, 2006).

2. The National Environmental Policy Act of 1970 is a notable exception, but its influence has been considerably less than its authors hoped. See Caldwell (1998).

3. See Kalinowski (unpublished manuscript).

4. Congressional politics has compounded the problems of executive power by rendering the body less effective than it ought to have been. Mann and Ornstein (2006) provide a useful analysis of the problem and possible solutions.

5. See Sunstein (2004).

6. Useful summaries of climate change effects are in Lynas (2007) and Walker and King (2008), pp. 53–86.

7. Walker and King, pp. 49–50.

8. See Broecker and Kunzig (2008), p. 138.

9. Battisti and Naylor's research (2009) indicates that present trends will similarly impact agriculture in tropical and subtropical regions worldwide. Average temperatures by the year 2100, they say, will exceed the most extreme temperatures recorded from 1900 to 2006.

10. Paul Epstein (2000); see also Center for Health and the Global Environment (2006).

11. The evidence is sizable; see for example Colin Campbell (2005), Darley (2004), David Goodstein (2004), Heinberg (2003), Klare (2004), Leggett (2005), Roberts (2004), and Simmons (2005). For a contrary view, see Smil (2008).

12. James Benson, Charles Chichetti, Herman Daly, Bruce Hannon, Denis Hayes, Amory Lovins, Eugene Odum, and Cecil Phillips.

13. Energy return on investment is a controversial subject, but the long-term trends for fossil fuels are less so. See Shah (2004), p. 161.

14. See Johnson (2000).

15. The Commission on the Prevention of Weapons of Mass Destruction, Proliferation, and Terrorism concludes that it is "more likely than not that a weapon of mass destruction will be used in a terrorist attack somewhere in the world by 2013" (2008, p. xv).

16. For analysis prior to the financial implosion of 2008, see Kuttner (2007) and Phillips (2008).

17. The best analyses are those by Makhijani (2007), Lovins et al. (2005), and Inslee and Hendricks (2007).

18. Reliable numbers here are hard to come by, partly because the subject is complex and partly because they would reveal inconvenient truths about the sanity of the energy system. A full accounting of the costs of the current energy system would include, for example, the effects of climate destabilization, the ecological effects of acid rain and mercury contamination, whatever price one might put on the tens of thousands who die early because they live downwind from coal plants, the costs of maintaining a military presence in the Middle East and periodically fighting oil wars that further raise the prospects of terrorist attacks in the United States, and subsidies for oil, gas, and nuclear power insinuated throughout federal and state budgets, as well as research and development expenditures heavily weighted to the status quo... and so forth.

19. There are cogent arguments and considerable evidence that renewables and radically improved efficiency are inadequate without, in Ted Trainer's words, a "radical change to a very different kind of society," one not organized around consumption (Trainer, 2007).

20. The debate about sequestering carbon is likely to continue indefinitely. The definitive study carried out by an MIT research team in 2007 is on the surface positive about the prospects, but hedges its bets in the fine print. The issue is whether carbon sequestration is feasible and can compete fairly with efficiency improvements and renewable energy.

21. The civil liberty implications of nuclear power have been mostly ignored in the current debate, but the analysis by Ayres (1975) is still cogent, amplified by the threat of terrorism. See also Lovins (forthcoming, 2009), a full analytical equivalent of a wooden stake through the heart of the ghoul. Cooke (2009, p. 407) describes the nuclear industry as a "huge, secretive, self-rationalizing system... backed by history, money, power, and a default conviction in its own inevitability." That outcome would have been no surprise to Dwight Eisenhower.

22. The case for a steady-state economy has been made thoroughly by Herman Daly, among others, and has been mostly dismissed by mainstream economists. This is both an interesting chapter in the history of ideas and a fairly ominous chapter in abnormal psychology. See Daly and Farley (2004) and Daly (1996).

23. U. Thara Srinivasan et al. (2008).

24. Victor (2008, p. 183) wisely cautions that "a 'no-growth' policy can be disastrous if implemented carelessly." Most likely, he thinks, no-growth will be driven from the grass roots (p. 222).

25. The CGIAR report is available at news.bbc.co.uk/2/hi/science/nature/6200114.

26. One of the most astute observers of American politics, Godfrey Hodgson, puts it this way: "The crucial change was the discrediting of government. This was possible because a substantial proportion of the American population, perturbed by the prospect of racial upheaval, rejected the ideals or the methods of the Great Society program. The methods may have been faulty, but the ideals were not, and in rejecting the methods, American society risked forgetting the ideals" (2004, p. 301).

27. Opinions about capitalism and its relation to the human prospect range widely. At one end of the spectrum, Klein (2007) paints a dismal picture of Milton Friedman and market fundamentalists and the mischief they've loosed on the world. Saul (2005) is slightly more encouraging, but not much. Esty and Winston (2006) and Hawken, Lovins, and Lovins (1998) are much more upbeat about possibilities for a green capitalism but dismissive of politics.

28. See Speth (2008), pp. 165–195. The recent writing on corporations and the environment has ranged from the breathy optimism about "triple bottom lines" to the "can't get there from here" variety. Noteworthy reading includes Bakkan (2004), Kelly (2001), Nace (2003), and Porritt (2006).

29. See Hartmann (2002), pp. 100–109.

30. Porritt (2006), 219–220.

31. But they had no such concern about the near stranglehold of corporations on our politics, the increasingly sinister intrusions of the military and paramilitary organizations, or the surveillance apparatus in our daily lives.

32. See John Ehrenfeld's useful discussion of "adaptive governance" (Ehrenfeld, 2008, pp. 182–196).

33. Pollan (2008).

34. The best description of resilience is still that of Lovins (1982), chapter 13; see also Murphy (2008).

Chapter 2

1. Webb (1952).

2. Ewen (1976), pp. 160–161; Ewen (1988), pp. 267–268; and Leach (1993), pp. 319–322.

3. For a good updating of Bernays, see Benjamin Barber's description of the infantilization of consumers in Barber (2007), pp. 81–115; Hamilton (2004) proposes beginning "by imposing restrictions on the quantity and nature of marketing messages, by first banning advertising and sponsorship from all public spaces and restricting advertising time on television and radio . . . tax laws could be changed so that costs of advertising are no longer a deductible business expense" (2004, p. 219).

4. Diamond (2005), Homer-Dixon (2006), Tainter (1989).

5. Hawken, Lovins, and Lovins (1999) and Esty and Winston (2006).

6. Again, Klein (2007) is instructive, as is Kitman (2000).

7. For a recent account see Beatty (2007), pp. 109–191.

8. Helen Thomas says "nothing is more troubling to me than the obsequious press during the run-up to the invasion of Iraq. They lapped up everything the Pentagon and the White House could dish out—no questions asked" (2006, p. 135).

9. Psychiatrist and neuroscientist Peter Whybrow (2005, p. 253) explains the emergence of decline of constraints on envy and greed as a "vicious cycle" that arises thus: "during times of great abundance, unless the prudence of frontal lobe reasoning imposes collective constraint through cultural agreement, human social behavior will run away to greed as the brain's ancient centers of instinctual self-preservation engage in a frenzy of self-reward."

10. See Weston (2008), Weston and Bach (2008), and other papers from the Climate Legacy Project at the Vermont Law School.

11. See the master's thesis of the same title by Jessica Boehland (2008).

Chapter 3

1. www.climateactionproject.com, and the published version, Becker (2009).

2. The study of leadership as practiced in America has been predominantly focused on business. Among the useful studies of leadership of a wider sort are those of historian James MacGregor Burns (1978) and Garry Wills (1994).

Chapter 4

1. Joy (2000); Sinsheimer (1978) made the same point.

2. See physicist Fritjof Capra's remarkable books *The Web of Life* (1996) and *The Hidden Connections* (2002); also Goerner, Dyck, and Lagerroos (2008).

Chapter 5

1. Simon et al. (2005), pp. 1689–1692; also my response in the same issue, pp. 1697–1698.

2. I make a distinction between evangelicals and extreme fundamentalists. My target here is exclusively the latter.

3. See George Marsden's description (Marsden, 2006, pp. 247–252).

4. Hoffer (1951) remains the classic description.

5. "First they came for the communists and I did not speak out because I was not a communist. Then they came for the trade unionists and I did not speak out because I was not a trade unionist. Then they came for the Jews and I did not speak out because I was not a Jew. Then they came for me and there was no one left to speak for me."

6. An early version is the 10th-century Islamic tale *The Case of the Animals Versus Man before the King of the Jinn*. The story has humans land on an island with a large number of animals. The humans begin to exploit the animals, who bring their grievances to the king of the jinn who also live on the island. The king rules that humans may control the animals but affirms that God is the protector of the animals. See Said and Funk (2003).

7. After writing this I came across Robert Emmon's fine book *Thanks: How the New Science of Gratitude Can Make You Happier* (2007).

8. "Technology is addictive. Material progress creates problems that are—or seem to be—soluble only by further progress. Again, the devil here is in the scale. . . ." (Wright, 2005, p. 7).

9. For those other practitioners who have forgotten the list, they are pride, greed, lust, anger, envy, sloth, and gluttony.

Chapter 6

1. James Lovelock (1998 and 2006) is among the very few to conjecture about how to convey the rudiments of science and civilization in a durable and usable form to those living on the other side of a collapsed civilization.

Chapter 7

1. See Norton (2004).
2. Read political philosopher Brian Berry's compelling case for justice in a greenhouse world, titled "Justice or Bust" (2005, pp. 260–273).
3. This phrase is adapted from Wells (1946). Wells wrote: "This world is at the end of its tether. The end of everything we call life is close at hand and cannot be evaded" (p. 1).

Chapter 8

1. The U.S. Supreme Court is apparently losing a large share of its international audience who find its decisions, perhaps, too ideological, aloof, formulaic, and remote from lived reality. See Liptak (2008).
2. It is worth studying the similarities between slavery and our use of fossil fuels as a matter of intergenerational law. See for example, David Orr, "2020: A Proposal," in Orr (2002), pp. 143–151, and Mouhot (2008).
3. Robert Ornstein and Paul Ehrlich once proposed the creation of a "foresight institute" charged with evaluation of long-term trends and their consequences: Ornstein and Ehrlich (1989).
4. With the leadership of Tony Cortese and his staff at Second Nature, hundreds of colleges and universities, including Oberlin, have responded to the challenge by signing commitments to move toward carbon neutrality and are taking steps to reduce the use of fossil fuels. Reaching the goal of carbon neutrality will be easier where sunlight and hydropower are abundant and more difficult in regions like our own that are highly dependent on coal. In any event, the case for moving rapidly toward levels of energy efficiency that lower carbon emissions includes lower costs as the price of fossil energy rises, resilience in the face of price shocks and supply interruptions, and the moral obligation not to damage the world in which our graduates and our children will live.

Sources

Ackerman, Bruce, and James Fishkin. *Deliberation Day*. New Haven: Yale University Press, 2004.

Adorno, T.W., et al. *The Authoritarian Personality*. New York: Harper and Row, 1950.

Alley, Richard B. "Abrupt Climate Change." *Scientific American*, November 2004, 62–69.

Allport, Gordon W. *The Nature of Prejudice*. Cambridge, Mass.: Addison-Wesley, 1954.

Alterman, Eric. *Kubuki Democracy*. New York: Nation Books, 2011.

Altemeyer, Bob. *The Authoritarian Specter*. Cambridge, Mass.: Harvard University Press, 1996.

Altemeyer, Bob. "Highly Dominating, Highly Authoritarian Personalities." *Journal of Social Psychology* 144 (2004): 422–425.

Alter, Jonathan. *The Defining Moment: FDR's Hundred Days and the Triumph of Hope*. New York: Simon & Schuster, 2006.

Anderson, Ray. *Mid-Course Correction: Toward a Sustainable Enterprise*. Atlanta: Peregrinzilla Press, 1998.

Andrews, Edmund L., "Greenspan Concedes Flaws in Deregulatory Approach." *New York Times* (October 24, 2008), pp. 1, 6.

Archer, David. *The Long Thaw: How Humans Are Changing the Next 100,000 Years of Earth's Climate*. Princeton: Princeton University Press, 2009.

Asch, Solomon E. "Opinions and Social Pressure." *Scientific American* 193 (1955): 31–35.

Ayres, Russell W. "Policing Plutonium: The Civil Liberties Fallout." *Harvard Civil Rights Civil Liberties Law Review* 10 (1975): 369–443.

Bagdikian, Ben H. *The Media Monopoly*. Boston: Beacon, 2000.

Bakkan, Joel. *The Corporation*. New York: The Free Press, 2004.

Balint, Peter, et al. *Wicked Environmental Problems.* Washington, D.C.: Island Press, 2011.

Barber, Benjamin. *Consumed*. New York: Norton, 2007.

Barnes, Peter. *Capitalism 3.0: A Guide to Reclaiming the Commons*. San Francisco: Berrett-Koehler, 2006.

Barnes, Peter, Robert Costanza, Paul Hawken, David W. Orr, Elinor Ostrom, Alvaro Umana, Oran Young, "Creating an Earth Atmospheric Trust," Letter to *Science*.

Barney, Gerald O. *The Global 2000 Report to the President.* Washington, D.C.: U. S. Government Printing Office, 1980.

Barnosky, Anthony D., *Heatstroke: Nature in an Age of Global Warming.* Washington, D.C.: Island Press, 2009.

Barry, Brian. *Why Social Justice Matters*. Cambridge, U.K.: Polity Press, 2005.

Bartels, Larry. *Unequal Democracy*. New York: Russell Sage Foundation, 2008.

Battisti, David S., and Rosamond L. Naylor. "Historical Warnings of Future Food Insecurity with Unprecedented Seasonal Heat." *Science* 323 (2009): 240–248.

Bawer, Bruce. *Stealing Jesus: How Fundamentalism Betrays Christianity*. New York: Crown, 1997.

Beatty, Jack. *Age of Betrayal: The Triumph of Money in America, 1865–1900.* New York: Knopf, 2007.

Becker, Bill. *The 100 Day Action Plan to Save the Planet: A Climate Crisis Solution for the 44th President*. New York: St. Martin's Griffin, 2008.

Bennett, W. Lance, Regina G. Lawrence, and Steven Livingston. *When the Press Fails: Political Power and the News Media from Iraq to Katrina*. Chicago: University of Chicago Press, 2007.

Benyus, Janine. *Biomimicry: Innovation Inspired by Nature*. New York: Morrow, 1997.

Berlin, Isaiah. *The Crooked Timber of Humanity: Chapters in the History of Ideas.* New York: Knopf, 1991.

Berry, Brian. *Why Social Justice Matters*. Cambridge, U.K.: Polity Press, 2005.

Berry, Thomas. *The Dream of the Earth*. San Francisco: Sierra Club Books, 1988.

Berry, Thomas. *The Great Work*. New York: Bell Tower, 1999.

Berry, Thomas. *Evening Thoughts*. San Francisco: Sierra Club Books, 2006.

Berry, Wendell. *Blessed Are the Peacemakers: Christ's Teachings of Love, Compassion & Forgiveness.* Washington: Shoemaker & Hoard, 2005.

Berry, Wendell. *The Memory of Old Jack*. New York: Harcourt Brace Jovanovich, 1974.

Black, Edwin. *Internal Combustion: How Corporations and Governments Addicted the World to Oil and Derailed the Alternatives*. New York: St. Martins, 2006.

Boehland, Jessica. "A Crime without a Name." Masters thesis, Yale University, School of Forestry, 2008.

Bollier, David. *Silent Theft: The Private Plunder of Our Common Wealth*. London: Routledge, 2003.

Boserup,Anders, and Andrew Mack. *War without Weapons: Non-violence in National Defence*. New York: Schocken, 1975.

Briggs, John Channing. *Lincoln's Speeches Reconsidered*. Baltimore: Johns Hopkins University Press, 2005.

Broecker, Wallace, and Robert Kunzig. *Fixing Climate*. New York: Hill & Wang, 2008.

Brown, Harrison. *The Human Future Revisited*. New York: Norton, 1978.

Brown, Lester R. *Plan B 3.0: Mobilizing to Save Civilization*. New York: Norton, 2008.

Brown, Peter G. *The Commonwealth of Life*. Montreal: Black Rose, 2008.

Brown, Peter G. *Restoring the Public Trust*. Boston: Beacon Press, 1994.

Burnell, Peter. *Climate Change and Democratization: A Complex Relationship*. (2009) Berlin: Heinrich Böll Foundation.

Burns, James MacGregor. *Leadership*. New York: Harpers, 1978.

Burns, James MacGregor. *Transforming Leadership*. New York: Grove Press, 2003.

Butler, George Lee. Remarks to the National Press Club, December 4, 1996. http://www.pbs.org/wgbh/amex/bomb/filmmore/reference/primary/leebutler.html

Butler, George Lee. "Zero Tolerance." *Bulletin of the Atomic Scientists*. 56 (2000): 20–21, 72–76.

Calaprice, Alice. *The New Quotable Einstein*. Princeton: Princeton University Press, 2005.

Caldwell, Lynton Keith. *The National Environmental Policy Act*. Bloomington: Indiana University Press, 1998.

Calvin, William. *Global Fever*. Chicago: University of Chicago Press, 2008.

Campbell, Colin. *Oil Crisis*. Essex, U.K.: Multi-Science Publishing, 2005.

Campbell, Kurt, ed. *Climatic Cataclysm: The Foreign Policy and National Security Implications of Climate Change*. Washington, D.C.: Brookings Institution, 2008.

Capra, Fritjof. *The Hidden Connections*. New York: Doubleday, 2002.

Capra, Fritjof. *The Web of Life*. New York: Anchor Books, 1996.

Catton, William. *Overshoot: The Ecological Basis of Revolutionary Change*. Urbana: University of Illinois Press, 1980.

Caudill, Harry. *Night Comes to the Cumberlands: A Biography of a Depressed Area*. Boston: Little, Brown, 1963.

Center for Health and the Global Environment. *Climate Change Futures: Health, Ecological and Economic Dimensions*. Cambridge, Mass.: Harvard Medical School, 2006. http://chge.med.harvard.edu/programs/ccf/index.html

Chameides, William, and Michael Oppenheimer. "Carbon Trading Over Taxes." *Science* 315 (2007): 1670.

Chardin, Teilhard de. *The Phenomenon of Man*. New York: Harper Torchbooks, 1965.

Cohen, Adam. *Nothing to Fear: FDR's Inner Circle and the Hundred Days that Created Modern America*. New York: Penguin, 2009.

Commission on the Prevention of Weapons of Mass Destruction, Proliferation, and Terrorism. *World at Risk*. New York: Vintage, 2008.

Connell, Evan. *Son of the Morning Star: Custer and the Little Bighorn*. New York: Harper Perennial, 1985.

Cooke, Stephanie. *In Mortal Hands: A Cautionary History of the Nuclear Age*. New York: Bloomsbury, 2009.

Council on Environmental Quality. *The Costs of Sprawl*. Washington: U.S. Government Printing Office, 1974.

Crow, Michael. "American Research Universities During the Long Twilight of the Stone Age." Remarks at the Rocky Mountain Sustainability Summit, February 21, 2007.

Crow, Michael. "None Dare Call It Hubris: The Limits of Knowledge." *Issues in Science and Technology*. Winter 2007, 1–4.

Dahl, Robert. *How Democratic Is the American Constitution?* New Haven: Yale University Press, 2002.

Dalai Lama, *Ethics for the New Millennium*. New York: Riverhead Books, 1999.

Daly, Herman. *Beyond Growth*. Boston: Beacon Press, 1996.

Daly, Herman. "On a Road to Disaster." *New Scientist*, October 18, 2008, 46–47.

Daly, Herman. *Steady-State Economics*. Washington, D.C.: Island Press, 1991.

Daly, Herman. *Valuing the Earth: Economics, Ecology, Ethics*. Cambridge, Mass.: MIT Press, 1996.

Daly, Herman, and Joshua Farley. *Ecological Economics*. Washington, D.C.: Island Press, 2004.

Damasio, Anthony. *Decartes' Error: Emotion, Reason, and the Human Brain*. New York: Grossett, 1994.

Darley, Julian. *High Noon for Natural Gas*. White River Junction, Vt.: Chelsea Green, 2004.

Dawkins, Richard. *The God Delusion*. Boston: Houghton Mifflin, 2006.

Dean, John. *Broken Government: How Republican Rule Destroyed the Legislative, Executive, and Judicial Branches*. New York: Viking, 2007.

Diamond, Jared. *Collapse: How Societies Choose to Fail or Succeed*. New York: Viking, 2005.

Douthwaite, Richard. *The Growth Illusion*. Tulsa: Council Oak Books, 1992.

Duncan, David James. "What Fundamentalists Need for Their Salvation." *Orion*. July/August 2005, 17–23.

Dyer, Gwynne. *Climate Wars*. Oxford: One World, 2010.

Eagan, David, et al. *Higher Education in a Warming World*. Washington, D.C.: National Wildlife Federation, 2008.

Ehrenfeld, David. *Becoming Good Ancestors*. New York: Oxford University Press, 2008.

Ehrenfeld, John. *Sustainability by Design*. New Haven: Yale University Press, 2008.

Ehrlich, Paul, and Anne Ehrlich, *The Dominant Animal: Human Evolution and the Environment*. Washington, D.C.: Island Press, 2008.

Eliot, T. S. *The Complete Poems and Plays*. New York: Harcourt, Brace & World, 1971.

Emanuel, Kerry. "Increasing Destructiveness of Tropical Cyclones over the Past 30 Years." *Nature* 436 (2005): 686–688.

Emmons, Robert. *Thanks! How the New Science of Gratitude Can Make You Happier*. Boston: Houghton Mifflin, 2007.

Epstein, Paul R. "Is Global Warming Harmful to Health?" *Scientific American*. August 2000, 50–57.

Epstein, Richard. *Supreme Neglect: How to Revive Constitutional Protection for Private Property*. New York: Oxford, 2008.

Epstein, Richard. *Takings: Private Property and the Power of Eminent Domain*. Cambridge, Mass.: Harvard University Press, 1985.

Erickson, Erik. *Childhood and Society*. New York: Norton, 1963.

Esty, Daniel, and Andrew Winston. *Green to Gold: How Smart Companies Use Environmental Strategy to Innovate, Create Value, and Build Competitive Advantage*. New Haven: Yale University Press, 2006.

Ewen, Stuart. *All Consuming Images: The Politics of Style in Contemporary Culture*. New York: Basic Books, 1988.

Ewen, Stuart. *Captains of Consciousness*. New York: McGraw-Hill, 1976.

Ferguson, Niall. *The Ascent of Money*. New York: Penguin Books, 2008.

Festinger, Leon. *A Theory of Cognitive Dissonance*. Evanston: Row, Peterson, 1957.

Fischetti, Mark. "Drowning New Orleans." *Scientific American*. October 2001, 76–85.

Flannery, Tim. *The Weather Makers: How Man Is Changing the Climate and What It Means for Life on Earth*. New York: Atlantic Monthly Press, 2006.

Frank, Thomas. *The Wrecking Crew: How Conservatives Rule*. New York: Metropolitan Books, 2008.

Frankl, Viktor. *Man's Search for Meaning*. London: Rider, 2004.

Freyfogle, Eric. *The Land We Share*. Washington, D.C.: Island Press, 2003.

Friedman, Benjamin. *The Moral Consequences of Economic Growth*. New York: Knopf, 2005.

Friedman, Thomas. *Hot, Flat, and Crowded*. New York: Farrar, Straus & Giroux, 2008.

Friedman, Thomas. "The Power of Green." *New York Times Magazine*, April 15, 2007.

Friedman, Thomas. "What Was That All About?" *New York Times*, December 18, 2007.

Fromm, Eric. *To Have or To Be*. New York: Bantam Books, 1981.

Frumkin, Howard, Lawrence Frank, and Richard Joseph Jackson. *Urban Sprawl and Public Health: Designing, Planning, and Building for Healthy Communities*. Washington, D.C.: Island Press, 2004.

Galbraith, James K. *The Predator State: How Conservatives Abandoned the Free Market and Why Liberals Should Too*. New York: The Free Press, 2008.

Galbraith, John Kenneth. *The Essential Galbraith*. New York: Mariner, 2001.

Gandhi, Mahatma. *The Essential Gandhi*. Edited by Louis Fischer. New York: Random House, 2002.

Gandhi, Mahatma. *Satyagraha in South Africa*. Stanford, Calif.: Academic Reprints, 1954.

Garvey, James. *The Ethics of Climate Change*. New York: Continuum, 2008.

Gelbspan, Ross. *Boiling Point: How Politicians, Big Oil and Coal, Journalists, and Activists Have Fueled a Climate Crisis—And What We Can Do to Avert Disaster*. New York: Basic Books, 2004.

Gelbspan, Ross. *The Heat Is On: The Climate Crisis, the Cover-up, the Prescription*. Reading, Mass.: Perseus Books, 1998.

Georgescu-Roegen, Nicholas. *The Entropy Law and the Economic Process: Tales from the Cruise Adventure of a Lifetime*. Cambridge, Mass.: Harvard University Press, 1971.

Goerner, Sally, Robert Dyck, and Dorothy Lagerroos. *The New Science of Sustainability*. Chapel Hill: Triangle Center for Complex Systems, 2008.

Goodstein, David, *Out of Gas: The End of the Age of Oil*. New York: Norton, 2004.

Goodstein, Eban. *Fighting for Love in the Century of Extinction*. Burlington: University of Vermont, 2007.

Gore, Al. "The Threat to American Democracy." Unpublished remarks at the Media Center, The Associated Press, New York, October 5, 2005.

Gore, Al. *The Assault on Reason*. New York: Penguin, 2007.

Gould, Kira, and Lance Hosey. *Women in Green: Voices of Sustainable Design*. Kansas City: Ecotone Publishing Co., 2007.

Grant, Lindsay. "Sustainability and Governmental Foresight." In *Global Survival,* edited by Ervin Laszlo and Peter Seidel,. New York: SelectBooks, 2006.

Greenleaf, Robert. *Servant Leadership*. Mahwah, N.J.: Paulist Press, 1977.

Gregg, Richard. *The Power of Nonviolence*. New York: Schocken, 1971.

Guelzo, Allen C. *Lincoln's Emancipation Proclamation: The End of Slavery in America*. New York: Simon & Schuster, 2004.

Gulledge, Jay. "Three Plausible Scenarios of Future Climate Change." In *Climatic Cataclysm*, 49–96, edited by Kurt Campbell, 2008.

Gutman, Amy, and Dennis Thompson. *Why Deliberative Democracy?* Princeton: Princeton University Press, 2004.

Hacker, Jacob S., and Paul Pierson. *Off Center: The Republican Revolution & the Erosion of American Democracy*. New Haven: Yale University Press, 2005.

Hallie, Philip. *Lest Innocent Blood Be Shed: The Story of the Village of Le Chambon and How Goodness Happened There*. New York: Harpers, 1994.

Halpin, John, et al. *The Structural Imbalance of Political Talk Radio*. Washington, D.C.: Center for American Progress, 2008.

Hamilton, Clive. *Growth Fetish*. London: Pluto Press, 2004.

Hamilton, Clive. *Requiem for a Species*. London: Earthscan, 2010.

Hansen, James. "Can We Still Avoid Dangerous Human-Made Climate Change?" Presentation at the New School University, New York City, February 10, 2006. http://www.columbia.edu/~jeh1/2006/NewSchool_20060210.pdf (accessed February 28, 2009).

Hansen, James. "Global Warming Twenty Years Later: Tipping Points Near." Presentation at National Press Club, Washington, D.C., June 23, 2008. http://www.columbia.edu/~jeh1/2008/TwentyYearsLater_20080623.pdf (accessed February 28, 2009).

Hansen, James. "Scientific Reticence and Sea Level Rise." *Environmental Research Letters* 2 (2007): 024002.

Hansen, James. *Storms of my Grandchildren*. New York: Bloomsbury, 2009.

Hansen, James. "Tipping Point." In *State of the Wild: 2008–2009*, edited by Eva Fearn. Washington, D.C.: Island Press, 2008.

Hansen, James, et al. "Climate Change and Trace Gases," *Philosophical Transactions of the Royal Society* 365 (2007): 1925–1954.

Hansen, James, et al. "Earth's Energy Imbalance." *Science* 308 (2005): 1431–1434.

Hansen, James, et al. "Global Temperature Change." *Proceedings of the National Academy of Sciences* 103 (2006): 14288–14293.

Hansen, James, et al., "Target Atmospheric CO_2: Where Should Humanity Aim?" *Open Atmospheric Science Journal* 2 (2008): 217–231.

Hardin, Garrett, *Exploring New Ethics for Survival: The Voyage of the Spaceship Beagle*. Baltimore: Penguin Books, 1972.

Hardin, Garrett. "The Tragedy of the Commons." *Science* 162 (1968): 1243–48.

Harris, Sam. *The End of Faith*. New York: Norton, 2004.

Hart, Gary. *The Shield and the Cloak*. New York: Oxford University Press, 2006.

Hartmann, Thom. *Unequal Protection: The Rise of Corporate Dominance and the Theft of Human Rights*. Emmaus: Rodale Press, 2002.

Havel, Vaclav. *Disturbing the Peace*. New York: Vintage, 1991.

Havel, Vaclav. *Living in Truth*. London: Faber and Faber, 1989.

Havel, Vaclav. *Summer Meditations*. New York: Knopf, 1992.

Havel, Vaclav. *The Art of the Impossible*. New York: Knopf, 1997.

Hawken, Paul, Amory Lovins, and Hunter Lovins. *Natural Capitalism: Creating the Next Industrial Revolution*. Boston: Little, Brown, 1999.

Heilbroner, Robert. *An Inquiry into the Human Prospect*. New York: Norton, 1980.

Heinberg, Richard. *The Party's Over*. Gabriola Island, Canada: New Society Publishers, 2003.

Heinberg, Richard. *Power Down: Options and Actions for a Post-carbon World*. Gabriola Island, Canada: New Society Publishers, 2004.

Heschel, Abraham. *Man Is Not Alone: A Philosophy of Religion*. New York: Farrar, Straus & Giroux, 1990.

Hill, Steven. *10 Steps to Repair American Democracy*. Sausalito, Calif.: PoliPointPress, 2006.

Hill, Steven. *Fixing Elections: The Failure of America's Winner Take All Politics.* New York: Routledge, 2002.

Hillman, Mayer, Tina Fawcett, and Sudhir Chella Rajan. *The Suicidal Planet: How to Prevent Global Climate Catastrophe.* New York: St. Martin's Press, 2007.

Hodgson, Godfrey. *More Equal Than Others: America from Nixon to the New Century.* Princeton: Princeton University Press, 2004.

Hoffer, Eric. *The True Believer.* New York: Harper and Row, 1951.

Holdren, John. "One Last Chance to Lead," *Scientific American Earth 3.0.* Fall 2008, 20–21.

Holdren, John. "Science and Technology for Sustainable Well-Being." *Science* 319 (2008): 424–434.

Holzer, Harold. *Lincoln at Cooper Union: The Speech That Made Abraham Lincoln President.* New York: Simon & Schuster, 2004.

Homer-Dixon, Thomas, and David Kieth. "Blocking the Sky to Save the Earth." *New York Times,* September 20, 2008.

Homer-Dixon, Thomas. *The Ingenuity Gap: Facing the Economic, Environmental, and Other Challenges of an Increasingly Complex and Unpredictable Future.* New York: Knopf, 2000.

Homer-Dixon, Thomas. *The Upside of Down: Catastrophe, Creativity, and the Renewal of Civilization.* Washington, D.C.: Island Press, 2006.

Hopkins, Rob. *The Transition Handbook.* White River Jct: Chelsea Green, 2008.

Horton, Scott. "Vote Machine." *Harper's Magazine.* March 2008, 37–46.

Horwitz, Morton. *The Transformation of American Law: 1780–1860.* Cambridge, Mass.: Harvard University Press, 1977.

Illich, Ivan. *Energy and Equity.* New York: Harper and Row, 1974.

Inslee, Jay, and Bracken Hendricks. *Apollo's Fire: Igniting America's Clean Energy Economy.* Washington, D.C.: Island Press, 2007.

Intergovernmental Panel on Climate Change. 3 vols. New York: Cambridge University Press, 2007.

Jackson, Tim. "What Politicians Dare Not Say." *New Scientist,* October 18, 2008, 42–43.

Jacoby, Susan. *The Age of American Unreason.* New York: Pantheon, 2008.

Janis, Irving. *Victims of Groupthink.* Boston: Houghton Mifflin, 1972.

Jarrett, James (ed.). *Nietzsche's Zarathustra: Notes of the Seminar Given in 1934–1939.* Princeton, N.J.: Princeton University Press, 1988.

Joffe, Paul. "The Dwindling Margin for Error: The Realist Perspective on Global Governance and Global Warming." *Rutgers Journal of Law & Public Policy.* 5 (2007): 89–176.

Johnson, Chalmers. *Blowback: The Costs and Consequences of American Empire.* New York: Henry Holt, 2000.

Johnson, Chalmers. "Going Bankrupt: Why the Debt Crisis Is Now the Greatest Threat to the American Republic." TomDispatch.com, January 22, 2008. http://www.tomdispatch.com/post/174884 (accessed February 28, 2009).

Johnson, Chalmers. *Nemesis: The Last Days of the American Republic*. New York: Henry Holt, 2006.

Jones, Van. *The Green Collar Economy: How One Solution Can Fix Our Two Biggest Problems*. New York: HarperOne, 2008.

Joy, Bill. "Why the Future Doesn't Need Us." *Wired*, April 2000.

Kalinowski, Frank. "An Ecological Interpretation of the Constitution." Unpublished manuscript.

Kaplan, Rachel, and Stephen Kaplan. *The Experience of Nature: A Psychological Perspective*. Cambridge, U.K.: Cambridge University Press, 1989.

Kaplan, Robert. "Was Democracy Just a Moment?" *The Atlantic Monthly*. December 1997, 55–80.

Katzer, James, et al. *The Future of Coal: An Interdisciplinary MIT Study*. Cambridge, Mass.: MIT Press, 2007.

Kauffman, Stuart. *Reinventing the Sacred: A New View of Science, Reason, and Religion*. New York: Basic Books, 2008.

Kellert, Stephen, Judith Heerwagen, and Martin Mador, eds. *Biophilic Design: The Theory, Science, and Practice of Bringing Buildings to Life*. New York: Wiley, 2008.

Kelly, Marjorie. *The Divine Right of Capital: Dethroning the Corporate Aristocracy*. San Francisco: Berrett-Koehler, 2001.

Kelman, Steven. "Why Public Ideas Matter." In *The Power of Public Ideas*, edited by Robert Reich. Cambridge, Mass.: Harvard University Press, 1990.

Kennedy, Robert, Jr. *Crimes against Nature: How George W. Bush and His Corporate Pals Are Plundering the Country and Hijacking Our Democracy*. New York: HarperCollins, 2004.

Kitman, Jamie. "The Secret History of Lead." *The Nation*, March 20, 2000, 11–44.

Klare, Michael. *Blood and Oil: The Dangers and Consequences of America's Growing Dependency on Imported Petroleum*. New York: Metropolitan Books, 2004.

Klein, Naomi. *The Shock Doctrine: The Rise of Disaster Capitalism*. New York: Metropolitan Books, 2007.

Kogan, Richard, et al. "The Long-Term Fiscal Outlook is Bleak: Restoring Fiscal Sustainability Will Require Major Changes to Programs, Revenues, and the Nation's Health Care System." Washington: Center on Budget and Policy Priorities, January 29, 2007. http://www.cbpp.org/1-29-07bud.pdf (accessed February 28, 2009).

Kohak, Erazim. *The Embers and the Stars*. Chicago: University of Chicago Press, 1984.

Kolbert, Elizabeth. *Field Notes from a Catastrophe: Man, Nature, and Climate Change*. New York: Bloomsbury, 2006.

Korten, David. "Only One Reason to Grant a Corporate Charter." Speech at Faneuil Hall, Boston, November 13, 2007. Available at http://www.commondreams.org/archive/2007/12/08/5710 (accessed February 28, 2009).

Kraybill, Donald, Steven M. Nolt, and David L. Weaver-Zercher. *Amish Grace: How Forgiveness Transcended Tragedy*. New York: John Wiley, 2007.

Krupp, Fred. *Earth: The Sequel*. New York: Norton, 2008.

Kuhn, Thomas. *The Structure of Scientific Revolutions*. Chicago: University of Chicago Press, 1963.

Kuntsler, James Howard. *The Long Emergency: Surviving the Converging Catastrophes of the Twenty-first Century*. New York: Atlantic Monthly Press, 2005.

Kunzig, Robert. "A Sunshade for Planet Earth." *Scientific American*. November 2008, 46–55.

Kurlansky, Mark. *Nonviolence: Twenty-Five Lessons from the History of a Dangerous Idea*. New York: Modern Library, 2006.

Kurzweil, Ray. *The Singularity Is Near: When Humans Transcend Biology*. New York: Viking, 2005.

Kutscher, Charles, ed. *Tackling Climate Change in the U.S.* Boulder, Colo.: American Solar Energy Association, 2007.

Kuttner, Robert. "Failures of Politics." *The American Prospect*. April 2006, 3.

Kuttner, Robert. *Obama's Challenge: America's Economic Crisis and the Power of a Transformative Presidency*. White River Junction, Vt.: Chelsea Green, 2008.

Kuttner, Robert. *The Squandering of America: How the Failure of Our Politics Undermines Our Prosperity*. New York: Knopf, 2007.

Lakoff, George. *Don't Think of an Elephant: Know Your Values and Frame the Debate*. White River Junction: Chelsea Green, 2004.

Layard, Richard. *Happiness: Lessons from a New Science*. New York: Penguin, 2005.

Lazare, Daniel. *The Frozen Republic: How the Constitution Is Paralyzing Democracy*. New York: Harcourt Brace & Co., 1996.

Lazarus, Richard. *The Making of Environmental Law*. Chicago: University of Chicago Press, 2004.

Leach, William. *Land of Desire: Merchants, Power, and the Rise of a New American Culture*. New York: Pantheon, 1993.

Lear, Jonathan. *Radical Hope: Ethics in the Face of Cultural Devastation*. Cambridge, Mass.: Harvard University Press, 2006.

LeDoux, Joseph. *The Emotional Brain: The Mysterious Underpinnings of Emotional Life*. New York: Simon & Schuster, 1996.

Leggett, Jeremy. *The Empty Tank: Oil, Gas, Hot Air, and the Coming Global Financial Catastrophe*. New York: Random House, 2005.

Leopold, Aldo. *A Sand County Almanac*. New York: Oxford University Press, 1949.

Levinson, Sanford. *Our Undemocratic Constitution: Where the Constitution Goes Wrong (And How We the People Can Correct It)*. New York: Oxford University Press, 2006.

Lifton, Robert Jay, and Eric Markusen. *The Genocidal Mentality*. New York: Basic Books, 1990.

Lincoln, Abraham. *Lincoln: Speeches and Writings, 1859–1865*. Edited by Don Fehrenbacher. New York: Library of America, 1989.

Lindblom, Charles. *The Market System: What It Is, How It Works, and What to Make of It.* New Haven: Yale University Press, 2001.

Linden, Eugene. "Antarctica." *Time*, April 14, 1997.

Linebaugh, Peter. *The Magna Carta Manifesto: Liberties and Commons for All.* Berkeley: University of California Press, 2008.

Linker, Damon. *The Theocons: Secular America under Siege.* New York: Anchor Books, 2007.

Liptak, Adam. "U.S. Court, a Longtime Beacon, Is Now Guiding Fewer Nations." *New York Times*, September 18, 2008.

Locke, John. *Two Treatises of Government.* New York: Mentor Books, 1965.

Louv, Richard. *Last Child in the Woods: Saving Our Children from Nature-Deficit Disorder.* Chapel Hill: Algonquin Books, 2005.

Lovelock, James. "A Book for All Seasons." *Science* 280 (1998), pp. 832–833.

Lovelock, James. *The Revenge of Gaia.* London: Penguin Books, 2006.

Lovelock, James. *The Vanishing Face of Gaia: A Final Warning.* New York: Basic Books, 2009.

Lovelock, James. *Gaia: A New Look at Life on Earth,* 1979.

Lovins, Amory, and Hunter Lovins. *Brittle Power.* Andover: Brick House, 1982.

Lovins, Amory, and Hunter Lovins. *Brittle Power: Energy Strategy for National Security.* Andover, Mass.: Brick House, 1982.

Lovins, Amory, et al. *Winning the Oil Endgame.* Snowmass, Colo.: Rocky Mountain Institute, 2005.

Lovins, Amory. "The Nuclear Illusion." *Ambio* (forthcoming, 2009).

Lovins, Amory. *Small is Profitable.* Snowmass: Rocky Mountain Institute, 2002.

Lynas, Mark. *Six Degrees: Our Future on a Hotter Planet.* New York: HarperPerennial, 2007.

Lynn, Barry. *End of the Line: The Rise and Coming Fall of the Global Corporation.* New York: Doubleday, 2005.

Macy, Joanna, and Jonathan Seed, *Thinking like a Mountain: Toward a Council of all Beings.* Philadelphia: New Society Press, 1988.

Makhijani, Arjun. *Carbon-Free and Nuclear-Free: A Roadmap for U.S. Energy Policy.* Takoma Park: IEER Books, 2007.

Mann, Thomas, and Norman Ornstein. *The Broken Branch: How Congress Is Failing America and How to Get It Back on Track.* New York: Oxford, 2006.

Mannheim, Karl. *Man and Society in an Age of Reconstruction.* New York: Harcourt, Brace & World, Inc., 1940.

Marsden, George. *Fundamentalism and American Culture.* New York: Oxford University Press, 2006.

Maslow, Abraham. *The Farther Reaches of Human Nature.* New York: Viking, 1971.

Maslow, Abraham. *The Psychology of Science.* Chicago: Gateway Books, 1966.

Matthews, Jessica Tuchman. "Redefining Security." *Foreign Affairs* (Spring, 1989).

Matthews, Richard K. *If Men Were Angels: James Madison and the Heartless Empire of Reason.* Lawrence: University of Kansas Press, 1995.

Mayer, Stephen, and Cindy Frantz. "The Connectedness to Nature Scale: A Measure of Individuals' Feeling in Community with Nature." *Journal of Environmental Psychology* 24 (2004): 503–515.

Mayer, Stephen, et al. "Why Is Nature Beneficial? The Role of Connectedness to Nature." *Environment and Behavior* (in press).

McChesney, Robert. *Rich Media, Poor Democracy.* Urbana: University of Illinois Press, 1999.

McIntosh, Alastair. *Hell and High Water: Climate Change, Hope and the Human Condition.* Edinburgh: Birlinn, 2008.

McKibben, Bill. *Eaarth.* New York: Times Books, 2010.

McKibben, Bill. *Hope, Human and Wild: True Stories of Living Lightly on the Earth.* Boston: Little Brown, 1995.

McKibben, Bill. "The Christian Paradox." *Harper's,* August 2005, 31–37.

McKie, Robin, "Scientists to Issue Stark Warning over Dramatic New Sea Level Figures." *The Guardian,* March 8, 2009. Posted on www.commondreams.org.

McKinsey & Company. *Reducing U.S. Greenhouse Gas Emissions: How Much at What Cost?* Report published November 2007. http://www.mckinsey.com/clientservice/ccsi/greenhousegas.asp (accessed March 1, 2009).

McNeill, J. R. *Something New Under the Sun: An Environmental History of the Twentieth-Century World.* New York: Norton, 2000.

Meadows, Donella H. "Places to Intervene in a System." *Whole Earth.* Winter 1997, 78–84.

Meadows, Donella H. *Thinking in Systems: A Primer.* White River Junction, Vt.: Chelsea Green, 2008.

Merton, Robert. *Social Theory and Social Structure.* New York: Free Press, 1968.

Miles, Jack. "Global Requiem: The Apocalyptic Moment in Religion, Science, and Art." *CrossCurrents* 50 (2000), http://www.crosscurrents.org/milesrequiem.htm.

Milgram, Stanley. *Obedience to Authority.* New York: Harper and Row, 1969.

Millennium Ecosystem Assessment. *Our Human Planet: Summary for Decision-makers.* 5 vols. Washington, D.C.: Island Press, 2005.

MIT Study Group. "The Future of Coal: Options for a Carbon-Constrained World." 2007. http://web.mit.edu/coal/ (accessed March 1, 2009).

Monbiot, George. *Heat: How to Stop the Planet from Burning.* Cambridge, Mass.: South End Press, 2007.

Mooney, Chris. "Climate Repair Made Simple." *Wired.* July 2008, 129–133.

Moore, Barrington. *Reflections on the Causes of Human Misery.* Boston: Beacon, 1972.

Moser, Bob. "The Crusaders." RollingStone.com, April 2005. http://www.rollingstone.com/politics/story/7235393/the_crusaders/ (accessed March 1, 2009).

Mouhot, Jean-François. "Historical and Contemporary Links and Parallels in Slave Ownership and Fossil Fuel Usage." Unpublished manuscript, 2008.

Murphy, Pat. *Plan C: Community Survival Strategies for Peak Oil and Climate Change.* Gabriola Island, B.C.: New Society Publishers, 2008.

Nace, Ted. *Gangs of America: The Rise of Corporate Power and the Disabling of Democracy*. San Francisco: Berrett-Koehler, 2003.

National Science and Technology Council. *Scientific Assessment of the Effects of Global Change on the United States*. Washington, D.C.: Office of the President, 2008. http://www.climatescience.gov/Library/scientific-assessment/

Neibuhr, Reinhold. *The Irony of American History*. New York: Scribners, 1952.

Nelson, Robert. "Rethinking the American Constitution." *Philosophy & Public Affairs Quarterly* 26 (2006): 2–11.

Norton, Anne. *Leo Strauss and the Politics of American Empire*. New Haven: Yale University Press, 2004.

Odum, Howard T., and Elisabeth C. Odum. *A Prosperous Way Down: Principles and Policies*. Boulder: University Press of Colorado, 2001.

Ophuls, William. *Ecology and the Politics of Scarcity Revisited*. San Francisco: W. H. Freeman, 1992.

Oreskes, Naomi, and Erik Conway. *Merchants of Doubt*. New York: Bloomsbury, 2010.

Ornstein, Robert, and Paul Ehrlich. *New World New Mind*. New York: Doubleday, 1989.

Orr, David W. "Security by Design," *Solutions* 3(1) (2012).

Orr, David W. "The Imminent Demise of the Republican Party." Common-Dreams.org, January 2005. http://www.commondreams.org/views05/0112-36.htm (accessed March 1, 2009).

Orr, David W. *The Nature of Design: Ecology, Culture, and Human Intention*. New York: Oxford University Press, 2002.

Orr, David W. "Speed." *Conservation Biology* 12 (1998): 4–7.

Orr, David W., and Stuart Hill. "Leviathan, the Open Society, and the Crisis of Ecology." *The Western Political Quarterly* 31 (1978): 457–469.

Overpeck, Jonathan, T. Bette, L. Otto-Bliesner, Gifford H. Miller, Daniel R. Muhs, Richard B. Alley, and Jeffrey T. Kiehl. "Paleoclimateic Evidence for Future Ice-Sheet Instability and Rapid Sea-Level Rise." *Science* 311 (2006): 1747–1750.

Pearce, Fred. *With Speed and Violence: Why Scientists Fear Tipping Points in Climate Change*. Boston: Beacon, 2007.

Perrow, Charles. *The Next Catastrophe: Reducing Our Vulnerabilities to Natural, Industrial, and Terrorist Disasters*. Princeton: Princeton University Press, 2007.

Phillips, Kevin. *Bad Money: Reckless Finance, Failed Politics, and the Global Crisis of American Capitalism*. New York: Viking, 2008.

Pittock, A. Barrie. "Ten Reasons Why Climate Change May Be More Severe Than Projected." In *Sudden and Disruptive Climate Change: Exploring the Real Risks and How We Can Avoid Them*, 11–27, edited by Michael C. MacCracken, Frances Moore, and John C. Topping Jr. London: Earthscan, 2008.

Podesta, John, and Peter Ogden. "Expected Climate Change over the Next Thirty Years." In *Climatic Cataclysm: The Foreign Policy and National Security*

Implications of Climate Change, 97–132, edited by Kurt Campbell. Washington: Brookings Institution, 2008.

Pollan, Michael. "Farmer in Chief." *New York Times Magazine,* October 12, 2008.

Porritt, Jonathon. *Capitalism as If the World Matters.* London: Earthscan, 2006.

Posner, Richard. *Catastrophe: Risk and Response.* New York: Oxford University Press, 2004.

Posner, Richard. *Law, Pragmatism, and Democracy.* Cambridge, Mass.: Harvard University Press, 2003.

Postman, Neil. *Building a Bridge to the Eighteenth Century: How the Past Can Improve Our Future.* New York: Knopf, 1999.

Powell, James Lawrence. *Dead Pool: Lake Powell, Global Warming, and the Future of Water in the West.* Berkeley: University of California Press, 2008.

Powell, John Wesley. *Report on the Lands of the Arid Region of the United States: With a More Detailed Account of the Lands of Utah.* Boston: The Harvard Common Press, 1983 (facsimile of 1879 ed).

Primack, Joel, and Nancy Abrams. 2006. *The View from the Center of the Universe: Discovering Our Extraordinary Place in the Cosmos.* New York: Riverhead Books, 2006.

Rabe, John. *The Good Man of Nanking: The Diaries of John Rabe.* New York: Vintage, 2000.

Ramo, Joshua Cooper. *The Age of the Unthinkable.* New York: Back Bay Books, 2010.

Raskin, Paul, et al. *The Great Transition: The Promise and Lure of the Times Ahead.* Boston: Stockholm Environment Institute, 2002.

Raupach, Michael R., et al. "Global and Regional Drivers of Accelerating CO_2 Emissions." *Proceedings of the National Academy of Sciences* 104 (2007): 10288–10293.

Reece, Erik. *Lost Mountain: A Year in the Vanishing Wilderness; Radical Strip Mining and the Devastation of Appalachia.* New York: Riverhead Books, 2006.

Rees, Martin. *Our Final Hour: A Scientist's Warning; How Terror, Error, and Environmental Disaster Threaten Humankind's Future in This Century—On Earth and Beyond.* New York: Basic Books, 2003.

Rees, Martin. "Science: The Coming Century." *New York Review of Books,* November 20, 2008, 41–44.

Repetto, Robert. "The Climate Crisis and the Adaptation Myth." Yale University School of Forestry and Environmental Studies, Working Paper 13 (2008). http://environment.research.yale.edu/documents/downloads/v-z/WorkingPaper13.pdf (accessed March 1, 2009).

Roberts, Paul. *The End of Oil: On the Edge of a Perilous New World.* Boston: Houghton Mifflin, 2004.

Robock, Alan. "20 Reasons Why Geoengineering May Be a Bad Idea." *Bulletin of the Atomic Scientists* 64 (2008): 14–18. http://www.thebulletin.org/files/064002006_0.pdf (accessed March 1, 2009).

Roddick, Anita, ed. *A Revolution in Kindness*. West Sussex, U.K.: Anita Roddick Books, 2003.

Rogers, Carl. *On Becoming a Person*. Boston: Houghton Mifflin, 1961.

Romm, Joseph. *Hell and High Water: Global Warming—the Solution and the Politics—and What We Should Do*. New York: William Morrow, 2007.

Rossing, Barbara. *The Rapture Exposed: The Message of Hope in the Book of Revelation*. New York: Basic Books, 2004.

Rossiter, Clinton. *Conservatism in America*. Cambridge, Mass.: Harvard University Press, 1982.

Roston, Eric. *The Carbon Age: How Life's Core Element Has Become Civilization's Greatest Threat*. New York: Walker & Co., 2008.

Roszak, Theodore. *The Voice of the Earth*. New York: Simon & Schuster, 1992.

Ruskin, John. *Unto This Last*. New York: Dutton, 1968.

Sabato, Larry. *A More Perfect Constitution: 23 Proposals to Revitalize Our Constitution and Make America a Fairer Country*. New York: Walker & Company, 2007.

Said, Abdul Aziz, and Nathan Funk. "Peace in Islam: An Ecology of the Spirit." In *Islam and Ecology: A Bestowed Trust*, edited by Richard G. Foltz, Frederick M. Denny, and Azizan Baharuddin. Cambridge, Mass.: Harvard University Press, 2003.

Samons, Loren, II. *What's Wrong with Democracy? From Athenian Practice to American Worship*. Berkeley: University of California Press, 2004.

Saul, John Ralston. *The Collapse of Globalism: And the Reinvention of the World*. London: Grove Atlantic, 2005.

Schaefer, Mark, et al. "An Earth Systems Science Agency." *Science* 321 (2008): 44–45.

Schell, Jonathan. *The Fate of the Earth*. Palo Alto: Stanford University Press, 2000.

Schell, Jonathan. *The Unconquerable World*. New York: Metropolitan Books, 2003.

Schlesinger, Arthur. *The Age of Roosevelt: The Coming of the New Deal*. Boston: Houghton Mifflin, 1965.

Schmookler, Andrew Bard. *The Parable of the Tribes*. Berkeley: University of California, 1984.

Schumacher, E. F. *A Guide for the Perplexed*. New York, Harper and Row, 1977.

Schumpeter, Joseph. *Capitalism, Socialism, and Democracy*. 3rd ed. New York: Harper Torchbooks, 1962.

Sears, Paul B. "Ecology—A Subversive Subject." *Bioscience* 14 (1964): 11–13.

Senge, Peter. *The Necessary Revolution: How Individuals and Organizations Are Working Together to Create a Sustainable World*. New York: Doubleday, 2008.

Sennett, Richard. *The Corrosion of Character: The Personal Consequences of Work in the New Capitalism*. New York: Norton: 1998.

Shah, Sonia. *Crude: The Story of Oil*. New York: Seven Stories Press, 2004.

Sharlet, Jeff. *The Family: The Secret Fundamentalism at the Heart of American Power*. New York: Harper, 2008.

Sharp, Gene. *The Politics of Nonviolent Action.* Boston: Porter Sargent, 1973.

Sharp, Gene. *Waging Nonviolent Struggle: 20th Century Practice and 21st Century Potential.* Boston: Porter Sargent, 2005.

Shenkman, Rick. *Just How Stupid Are We? Facing the Truth About the American Voter.* New York: Basic Books, 2008.

Shepard, Paul, and Daniel McKinley, eds. *The Subversive Science: Essays Toward an Ecology of Man.* Boston: Houghton Mifflin, 1969.

Shuman, Michael H., and Hal Harvey. *Security without War: A Post–Cold War Foreign Policy.* Boulder: Westview, 1993.

Sider, Ronald. *The Scandal of Evangelical Politics: Why Are Christians Missing the Chance to Really Change the World?* Grand Rapids: Baker Books, 2008.

Simmons, Matthew. *Twilight in the Desert: The Coming Saudi Oil Shock and the World Economy.* New York: John Wiley, 2005.

Simon, Stuart, et al. "Conservation Theology for Conservation Biologists—A Reply to David Orr." *Conservation Biology* 19 (2005): 1689–1692.

Sinsheimer, Robert. "The Presumptions of Science." *Daedalus* 107 (1978): 23–35.

Smil, Vaclav. *Global Catastrophes and Trends: The Next Fifty Years.* Cambridge, Mass.: MIT Press, 2008.

Smil, Vaclav. *Oil: A Beginner's Guide.* Oxford: Oneworld Books, 2008.

Smith, Jeffrey. "The Dissenter." *Washington Post Magazine.* December 7, 1997.

Smith, Joel B., Stephen H. Schneider, Michael Oppenheimer, Gary W. Yohe, William Hare, Michael D. Masstranrea, Anand Patwardhan, Ian Burton, Jan Corfee-Morlot, Chris H. D. Magadra, Hans-Martin Fussel, A. Barrie Pittock, Atiq Rahman, Avelino Suarez, and Jean-Pascal van Ypersele. "Assessing Dangerous Climate Change through an Update of the Intergovernmental Panel on Climate Change (IPCC) 'Reasons for Concern.'" *Proceedings of the National Academy of Science.* 2009. www.pnas.org/cgi/doi/10.1073/pnas.0812355106.

Snyder, Gary. *Back on the Fire.* Berkeley: Counterpoint, 2007.

Snyder, Gary. *Earth House Hold.* New York: New Directions, 1969.

Solnit, Rebecca. *Hope in the Dark: Untold Histories, Wild Possibilities.* New York: Nation Books, 2004.

Solomon, Susan, et al. "Irreversible Climate Change Due to Carbon Dioxide Emissions." *Proceedings of the National Academy of Sciences* 106 (2009): 1704–1709.

Speth, James Gustave. *The Bridge at the End of the World: Capitalism, the Environment, and Crossing from Crisis to Sustainability.* New Haven: Yale University Press, 2008.

Speth, James Gustave. *Red Sky at Morning: America and the Crisis of the Global Environment.* New Haven: Yale University Press, 2004.

Srinivasan, U. Thara, et al. "The Debt of Nations and the Distribution of Ecological Impacts from Human Activities." *Proceedings of the National Academy of Sciences* 105 (2008): 1768–1773.

Steffen, Will, et al. "Abrupt Changes: The Achilles' Heels of the Earth System." *Environment* 46 (2004): 8–20.

Steffen, Will, et al. *Global Change and the Earth System: A Planet under Pressure.* Berlin: Springer, 2004.

Steindl-Rast, David. *Gratefulness, the Heart of Prayer.* Mahwah, N.J.: Paulist Press, 1984.

Sterman, John. "Risk Communication on Climate." *Science* 322 (2008): 532–533.

Stern, Nicholas. *The Economics of Climate Change: The Stern Review.* Cambridge. U.K.: Cambridge University Press, 2007.

Stiglitz, Joseph. "Capitalist Fools." *Vanity Fair.* January 2009.

Stone, Christopher. *Should Trees Have Standing? And Other Essays on Law, Morals and the Environment.* Los Altos: William Kaufmann, 1974.

Stringfellow, William. *An Ethic for Christians & Other Aliens in a Strange Land.* Waco: Word Books, 1973.

Sullivan, William. "Nature at Home." In *Urban Place*, 237–252, edited by Patricia Barlett. Cambridge, Mass.: MIT Press, 2005.

Sunstein, Cass. *The Second Bill of Rights: FDR's Unfinished Revolution and Why We Need It More Than Ever.* New York: Basic Books, 2004.

Surowiecki, James. *The Wisdom of Crowds: Why the Many Are Smarter Than the Few and How Collective Wisdom Shapes Business, Economies, Societies, and Nations.* New York: Doubleday, 2004.

Swimme, Brian, and Thomas Berry. *The Universe Story: From the Primordial Flaring Forth to the Ecozoic Era—A Celebration of the Unfolding of the Cosmos.* New York: HarperCollins, 1992.

Tainter, Joseph. *The Collapse of Complex Societies.* Cambridge, U.K.: Cambridge University Press, 1989.

Taleb. *Nassim Nicholas, The Black Swan.* New York: Random House, 2010.

Taleb, Nassim. *The Black Swan: The Impact of the Highly Improbable.* New York: Penguin, 2007.

Tasch, Woody. *Inquiries into the Nature of Slow Money: Investing as If Food, Farms, and Fertility Mattered.* White River Junction, Vt.: Chelsea Green, 2008.

Tavris, Carol, and Elliot Aronson. *Mistakes Were Made (But Not by Me): Why We Justify Foolish Beliefs, Bad Decisions, and Hurtful Acts.* New York: Harcourt, 2007.

Teicher, Martin. "Scars that Won't Heal." *Scientific American.* March 2002, 68–75.

Tenner, Edward. *Why Things Bite Back: Technology and the Revenge of Unintended Consequences.* New York: Knopf, 1996.

The President's Climate Action Project. November 2008. www.climateaction-project.com (accessed January 2011).

Thomas, Helen. *Watchdogs of Democracy? The Waning Washington Press Corps and How It Has Failed the Public.* New York: Scribner, 2006.

Tidwell, Mike. *Bayou Farewell: The Rich Life and Tragic Death of Louisiana's Cajun Coast.* New York: Vintage, 2003.

Todorov, Tzvetan. *Facing the Extreme: Moral Life in the Concentration Camps.* New York: Metropolitan Books, 1996.

Trainer, Ted. *Renewable Energy Cannot Sustain a Consumer Society.* Dordrecht: Springer, 2007.

Trenberth, Kevin E. "Warmer Oceans, Stronger Hurricanes." *Scientific American* (July 2007), pp. 45–51.

Unamuno, Miguel de. *The Tragic Sense of Life in Men and Nations.* Princeton: Princeton University Press, 1977.

Union of Concerned Scientists. *World Scientists' Warning to Humanity.* Boston, 1992. http://www.ucsusa.org/about/1992-world-scientists.html.

United States Climate Change Science Program. *Coastal Sensitivity to Sea Level Rise: A Focus on the Mid-Atlantic Region.* Washington, D. C., 2009.

Vernadsky, Vladimir. *The Biosphere.* New York: Springer, 1997.

Victor, Peter. *Managing Without Growth: Slower by Design, Not Disaster.* Northampton, Mass.: Edward Elgar, 2008.

Walker, Brian, and David Salt. *Resilience Thinking.* Washington, D.C.: Island Press, 2006.

Walker, Gavrielle, and Sir David King. *The Hot Topic: What We Can Do About Global Warming.* London: Bloomsbury, 2008.

Wallis, Jim. "God's Own Party?" Truthout.org, May 12, 2005.

Weart, Spencer. "Global Warming: How Skepticism became denial," *Bulletin of the Atomic Scientists* 67(1)(2011):41–50.

Weart, Spencer. *The Discovery of Global Warming.* Cambridge, Mass.: Harvard University Press, 2003.

Weaver, Richard M. *Ideas Have Consequences.* Chicago: University of Chicago Press, 1984.

Webb, Walter Prescott. *The Great Frontier.* Austin: University of Texas Press, 1952/1975.

Webster, P. J., G. J. Holland, J. A. Curry, and H.-R. Chang. "Changes in Tropical Cyclone Number, Duration, and Intensity in a Warming Environment." *Science* 309 (2005): 1844–1846.

Weisman, Alan. *The World Without Us.* New York: St. Martin's, 2007.

Wells, H. G. *Mind at the End of Its Tether.* New York: Didier, 1946.

Weston, Burns. "Climate Change and Intergenerational Justice." *Vermont Journal of International Law* 9 (2008): 375–430.

Weston, Burns, and Tracy Bach. "Climate Change and Intergenerational Justice: 'Present Law, Future Law.'" Unpublished, 2008.

White, Ronald C., Jr. *The Eloquent President: A Portrait of Lincoln Through His Words.* New York: Random House, 2005.

White, Ronald C., Jr. *Lincoln's Greatest Speech: The Second Inaugural.* New York: Simon & Schuster, 2002.

Whybrow, Peter C. *American Mania: When More Is Not Enough.* New York: Norton, 2005.

Wiesenthal, Simon. *The Sunflower: On the Possibilities and Limits of Forgiveness.* New York: Schocken Books, 1997.

Wills, Garry. *Certain Trumpets: The Nature of Leadership.* New York: Simon & Schuster, 1994.

Wills, Garry. *Lincoln at Gettysburg: The Words that Remade America.* New York: Simon & Schuster, 1992.

Wills, Garry. *A Necessary Evil: A History of American Distrust of Government.* New York: Simon & Schuster, 1999.

Wilson, Edward O. *The Future of Life.* New York: Knopf, 2002.

Wilson, Edward O. *In Search of Nature.* Washington: Island Press, 1996.

Wink, Walter. *Engaging the Powers: Discernment and Resistance in a World of Domination.* Minneapolis: Fortress Press, 1992.

Wink, Walter. *Naming the Powers: The Language of Power in the New Testament.* Minneapolis: Fortress Press, 1984.

Wink, Walter. *Unmasking the Powers.* Minneapolis: Fortress Press, 1986.

Winner, Langdon. "Do Artifacts Have Politics?" Chap. 2 in *The Whale and the Reactor: A Search for Limits in an Age of High Technology.* Chicago: University of Chicago Press, 1986.

Wolin, Sheldon. *Democracy Incorporated: Managed Democracy and the Specter of Inverted Totalitarianism.* Princeton: Princeton University Press, 2008.

Woodwell, George. "Biotic Feedbacks from the Warming of the Earth." In *Biotic Feedbacks in the Global Climatic System,* 3–21, edited by George Woodwell and Fred Mackenzie. New York: Oxford University Press, 1995.

Wright, Ronald. *A Short History of Progress.* New York: Carroll & Graf, 2005.

Yarrow, Andrew. *Forgive Us Our Debts: The Intergenerational Dangers of Fiscal Irresponsibility.* New Haven: Yale University Press, 2008.

Zakaria, Fareed. "The Case for a Global Carbon Tax." *Newsweek,* April 16, 2007, 94–96.

Index

258 ⁓ INDEX

climate destabilization response of,
2, 4–7
communication substance and style
in, 99–100
Congressional, executive power
and, 222n4
corporations influencing, 225n31
Democrats in, 68, 75
Enlightenment influencing, 205
money in, 104
presidential commission for
changes in, 207–9
reform of, 177–78
Republicans in, 67–68, 75, 127–28,
130–31
transformation of, 205
population size, 26, 157
Porritt, Jonathon, 36–37
positive feedbacks, 20
Posner, Richard, 69–70
post-carbon prosperity, 212, 214
posterity
in Constitution, 72
defenders of, 75
protection of, 68, 71–76
Postman, Neil, 171
post–World War II world, 193, 196
Potter, David, 71
powerdown, 24
president
as educator-in-chief, 211
executive power of, 15–16, 94–95,
222n4
Presidential Climate Action Project,
206
presidential commission, for
governance, politics, and law
changes, 207–9
press. *See also* media
freedom of, 61, 190
Iraq invasion and, 61, 225n8
Primack, Joel, 145

procrastination penalty, 19
productive capital, 59
progress trap, 148
property law, 44–48
psychology
codes of conduct and, 170–71
limitations of, 167–68
public airwaves, 100, 178, 210
public awareness
of climate destabilization, 5, 9,
100
gradualism in, 187–88
happy news and, 185–86
leadership and, 189–90
level of, 21
panic and, 187
responsibility of, 7
shared national agenda and, 71
technical knowledge needed for,
192
public order
governance and, 39–42
testing of, 54

Rabe, John, 169
radical hope, 173
radio, 62–63
Rawls, John, 145
Reagan, Ronald, 108, 131
realism
of change's scale and pace,
198–99
diluted, 186, 188
idealism *v.*, 196
neoconservative, 199
reciprocity, 150
Rees, Martin, 1
*Reflections on the Causes of Human
Misery* (Moore), 162
refugees, climate, xi, 19–20, 32
Regional Greenhouse Gas Initiative,
188
religion. *See also* Christianity

Buddhism, 180, 201
dialogue across, 134
dispensation of, 133–34
extremists in, 134
Lincoln's use of, 89–90
renewable energy technologies, 104,
150, 188, 223n19
Republicans
Christians and, 127–28, 130–31
politics of, 67–68, 75, 127–28,
130–31
resilience
creation of, 42, 56
improvement of, 174–75
Revelation, 129, 132
revolution
climate destabilization and, 206
design, 56–57
kindness, 180
Robinson, Beatrice, 210
Roddick, Anita, 180
Rogers, Carl, 166
Roosevelt, Franklin, 17, 66, 92–94,
99, 104
Roosevelt, Theodore, 96, 100, 108, 210
Rossiter, Clinton, 65–66
Roszak, Theodore, 169

Sabato, Larry, 207
Safina, Carl, 137
Samons, Loren, 51
A Sand County Almanac (Leopold), 76
Santa Clara County v. Southern Pacific
Railroad, 208
Schell, Jonathan, 73, 194
Schlesinger, Arthur, 93
Schneider, Stephen, 218
Schumacher, E. F., 194
Schumpeter, Joseph, 51, 69
science, larger context of, 137
Scofield Reference Bible, 129
sea levels, rising, 18–19, 32, 44, 113
Sebald, W. G., 111

Second Nature, 227n4
Seed, Jonathan, 138
self-fulfilling prophecies, 167
self-inflicted extinction, 73
self-reflection, 163
Senge, Peter, 38, 41
September 11, 2001, 15–16
sequestration, carbon, 101, 224n20
Shakespeare, William, 148–49
Sharlet, Jeff, 129–30
Sharp, Gene, 201
Shenkman, Rick, 51
shifting baselines, 34
"Should Trees Have Standing?"
(Stone), 208
Shuman, Michael, 201
Sider, Ronald, 128
Silent Spring (Carson), 79
Simon, Stuart, 127
Sinsheimer, Robert, x
slavery
fossil fuels and, 227n2
framing of, 85, 87–92
Lincoln and, 84–89
slow economy, 80–81
Smith, Adam, xvi
Snyder, Gary, 48, 160–61
societal collapse, 54–55
Soddy, Frederick, 82–83
Solow, Robert, 105
Sophocles, 124
Spadaro, Jack, 113–16
sprawling development, 43–44, 106
steady-state economy, 30, 224n22
Steinbeck, John, 181
Steindl-Rast, Brother David, 149
Stern, Nicholas, 19
Stiglitz, Joseph, 56, 195
Stone, Christopher, 208
Strauss, Leo, 187
Stringfellow, William, 133
Supreme Court, 208, 227n1
Surowiecki, James, 60